Bounded and Almost Periodic Solutions of
Nonlinear Operator Differential Equations

Mathematics and Its Applications (*Soviet Series*)

Volume 55

Bounded and Almost Periodic Solutions of Nonlinear Operator Differential Equations

by

A. A. PANKOV
Institute of Applied Problems in Mechanics and Mathematics,
Ukrainian Academy of Sciences,
L'vov, U.S.S.R.

Springer-Science+Business Media, B.V.

Library of Congress Cataloging-in-Publication Data

Pankov, A. A. (Aleksandr Andreevich)
[Orgranichennye i pochti periodicheskie reshenii͡a nelineĭnykh different͡sial'no-operatornykh uravneniĭ. English]
Bounded and almost periodic solutions of nonlinear operator differential equations / by A.A. Pankov ; [translated from the Russian by V.S. Zajackovski and A.A. Pankov].
p. cm. -- (Mathematics and its applications (Soviet series) ; v. 55)
Expanded and revised translation of: Ogranichennye i pochti periodicheskie reshenii͡a nelineĭnykh different͡sial'no-operatornykh uravneniĭ.
Includes bibliographical references.
ISBN 0-7923-0585-X
1. Operator equations--Numerical solutions. 2. Differential equations, Non-linear--Numerical solutions. I. Title. II. Series: Mathematics and its applications (Kluwer Academic Publishers). Soviet series ; v. 55.
QA329.4.P3613 1990
515'.355--dc20 89-48125

ISBN 978-94-011-9684-0 ISBN 978-94-011-9682-6 (eBook)
DOI 10.1007/978-94-011-9682-6

Printed on acid-free paper

This is the expanded and revised translation of the original work
ОГРАНИЧЕННЫЕ И ПОЧТИ ПЕРИОДИЧЕСКИЕ РЕШЕНИЯ НЕЛИНЕЙНЫХ ДИФФЕРЕНЦИАЛЬНО-ОПЕРАТОРНЫХ УРАВНЕНИЙ
Published by Naukova Dumka, Kiev, © 1985.

Translated from the Russian by V.S. Zajackovski and A.A. Pankov

To Tanya

SERIES EDITOR'S PREFACE

'Et moi, ..., si j'avait su comment en revenir, je n'y serais point allé.'

Jules Verne

The series is divergent; therefore we may be able to do something with it.

O. Heaviside

One service mathematics has rendered the human race. It has put common sense back where it belongs, on the topmost shelf next to the dusty canister labelled 'discarded non-sense'.

Eric T. Bell

Mathematics is a tool for thought. A highly necessary tool in a world where both feedback and non-linearities abound. Similarly, all kinds of parts of mathematics serve as tools for other parts and for other sciences.

Applying a simple rewriting rule to the quote on the right above one finds such statements as: 'One service topology has rendered mathematical physics ...'; 'One service logic has rendered computer science ...'; 'One service category theory has rendered mathematics ...'. All arguably true. And all statements obtainable this way form part of the raison d'être of this series.

This series, *Mathematics and Its Applications*, started in 1977. Now that over one hundred volumes have appeared it seems opportune to reexamine its scope. At the time I wrote

> "Growing specialization and diversification have brought a host of monographs and textbooks on increasingly specialized topics. However, the 'tree' of knowledge of mathematics and related fields does not grow only by putting forth new branches. It also happens, quite often in fact, that branches which were thought to be completely disparate are suddenly seen to be related. Further, the kind and level of sophistication of mathematics applied in various sciences has changed drastically in recent years: measure theory is used (non-trivially) in regional and theoretical economics; algebraic geometry interacts with physics; the Minkowsky lemma, coding theory and the structure of water meet one another in packing and covering theory; quantum fields, crystal defects and mathematical programming profit from homotopy theory; Lie algebras are relevant to filtering; and prediction and electrical engineering can use Stein spaces. And in addition to this there are such new emerging subdisciplines as 'experimental mathematics', 'CFD', 'completely integrable systems', 'chaos, synergetics and large-scale order', which are almost impossible to fit into the existing classification schemes. They draw upon widely different sections of mathematics."

By and large, all this still applies today. It is still true that at first sight mathematics seems rather fragmented and that to find, see, and exploit the deeper underlying interrelations more effort is needed and so are books that can help mathematicians and scientists do so. Accordingly MIA will continue to try to make such books available.

If anything, the description I gave in 1977 is now an understatement. To the examples of interaction areas one should add string theory where Riemann surfaces, algebraic geometry, modular functions, knots, quantum field theory, Kac-Moody algebras, monstrous moonshine (and more) all come together. And to the examples of things which can be usefully applied let me add the topic 'finite geometry'; a combination of words which sounds like it might not even exist, let alone be applicable. And yet it is being applied: to statistics via designs, to radar/sonar detection arrays (via finite projective planes), and to bus connections of VLSI chips (via difference sets). There seems to be no part of (so-called pure) mathematics that is not in immediate danger of being applied. And, accordingly, the applied mathematician needs to be aware of much more. Besides analysis and numerics, the traditional workhorses, he may need all kinds of combinatorics, algebra, probability, and so on.

In addition, the applied scientist needs to cope increasingly with the nonlinear world and the extra mathematical sophistication that this requires. For that is where the rewards are. Linear models are honest and a bit sad and depressing: proportional efforts and results. It is in the non-linear world that infinitesimal inputs may result in macroscopic outputs (or vice versa). To appreciate what I am hinting at: if electronics were linear we would have no fun with transistors and computers; we would have no TV; in fact you would not be reading these lines.

There is also no safety in ignoring such outlandish things as nonstandard analysis, superspace and anticommuting integration, p-adic and ultrametric space. All three have applications in both electrical engineering and physics. Once, complex numbers were equally outlandish, but they frequently proved the shortest path between 'real' results. Similarly, the first two topics named have already provided a number of 'wormhole' paths. There is no telling where all this is leading - fortunately.

Thus the original scope of the series, which for various (sound) reasons now comprises five subseries: white (Japan), yellow (China), red (USSR), blue (Eastern Europe), and green (everything else), still applies. It has been enlarged a bit to include books treating of the tools from one subdiscipline which are used in others. Thus the series still aims at books dealing with:

- a central concept which plays an important role in several different mathematical and/or scientific specialization areas;
- new applications of the results and ideas from one area of scientific endeavour into another;
- influences which the results, problems and concepts of one field of enquiry have, and have had, on the development of another.

The theory of bounded, periodic, and almost periodic solutions of (systems of) ordinary differential equations has a venerable history and a large literature. The present volume is concerned with the same kinds of solutions but now for the infinite dimensional case such as equations with retarded arguments, variational inequalities and evolution equations of the form $u_t + A(t)u + L(t)u = f$ where $L(t)$ is a linear unbounded operator and $A(t)$ is a monotone nonlinear operator. It is in this sense that the word operator in the title must be understood.

It is a pleasure to welcome in this series a book on this important topic by an author who has made substantial contributions to the subject (for instance to almost periodic solutions for a general class of abstract first order nonlinear evolution equations, nonlinear partial differential equations and to certain questions of Lions concerning variational inequalities).

The shortest path between two truths in the real domain passes through the complex domain.

J. Hadamard

La physique ne nous donne pas seulement l'occasion de résoudre des problèmes ... elle nous fait pressentir la solution.

H. Poincaré

Never lend books, for no one ever returns them; the only books I have in my library are books that other folk have lent me.

Anatole France

The function of an expert is not to be more right than other people, but to be wrong for more sophisticated reasons.

David Butler

Bussum, August 1990 Michiel Hazewinkel

Contents

INTRODUCTION

The theory of almost periodic (a.p.) functions was initiated between 1924 and 1926. Subsequently, important contributions were made by A. Besicovitch, S. Bochner, J. von Neumann, V. Stepanov, and B. Levitan. We especially mention those of von Neumann, who was the first to consider a.p. functions on groups.

From the earliest days on, the theory of a.p. functions has been connected with problems of differential equations. (Moreover, important work by P. Bohl and E. Esclangon had appeared before Bohr's contributions.) In his fundamental paper [126], J. Favard had introduced the notion of envelope of an a.p. differential equation (in modern interpretation, it is equivalent to considering a family of equations parametrised by the points of the Bohr compactification) and had connected the problem of existence of a.p. solutions with some separation properties of bounded solutions of equations which belong to the envelope. This work of Favard is the starting point of many further investigations.

In the present text, we are not concerned with the theory of ordinary (finite-dimensional) a.p. differential equations (see, for example, [35, 41, 127] and the references given there). Instead we fix our attention to infinite-dimensional evolution equations, both abstract and involving partial derivatives. Some important results for the linear case were obtained by Z. Amerio [100]. Starting from the well-known results of S. Sobolev [78] on the homogeneous wave equation, he investigated the corresponding inhomogeneous equation. Further, using the notion of weakly a.p. function introduced by him in [99], Amerio justified an infinite-dimensional version of the minimax method of constructing a.p. solutions (due to Favard) and obtained some generalizations of Favard's classical theorems. The strongest results in this direction were proved by V. Zhikov. These are presented systematically in [25, 41]. In the constant coefficient case an interesting approach was suggested in [31a].

We also note that recently the theory of linear partial differential equations with coefficients that are a.p. in all variables has gained interest (see, for example, [91, 94]).

Among the nonlinear a.p. evolution equations the class best understood is that of the abstract parabolic equations. For such equations G. Prouse [145] has obtained an existence theorem for a.p. solutions. Subsequently this result was strengthened considerably by V. Zhikov [22] (see also [25, 41]).

A new important class of evolution problems, known as variational inequalities, was introduced by J.-L. Lions and G. Stampaccia [140]. In 1969 Lions posed the question about bounded and a.p. solutions of such problems (see [44], problems

10.18 and 10.20 in Ch. 4). He also obtained a partial result in this direction (loc. cit., Theorem 8.4 of Ch. 4). M. Biroli [107-111] tried to answer the question of Lions and proved some results on a.p. solutions of variational inequalities. However, the existence theorems he obtained are not satisfactory, because in the case of parabolic equations they are weaker than those of Prouse. Later the problem was investigated in detail by the author [54, 57, 60, 67], and Lion's questions were answered completely.

Other classes of a.p. equations have also been investigated. For example, G. Prouse [144] proved almost periodicity of small solutions of the two-dimensional Navier-Stokes equations, and M. Brioli [12] extended this results to certain Navier-Stokes inequalities. There also many papers on nonlinear wave equations. G. Prouse [100] investigated a.p. solutions of a wave equation with nonlinear dissipative term. More general results were obtained by V. Zhikov [25]. Detailed investigation of dissipative wave equations can be found in [31a]. A.p. solutions of wave equations with small nondissipative nonlinearity were studied in [48, 49].

From the point of view of almost periodicity, a general class of first order abstract nonlinear evolution equations was investigated by the author in [66, 67]. This class contains the positive boundary value problems for symmetric hyperbolic systems with monotone nonlinearity and certain nonlinear Schrödinger equations.

As for nonlinear partial differential equations which are a.p. in the space variables, it seems that the first paper in this direction is that of M.A. Shubin [93]. This paper contains a generalization of Favard's first theorem to such problems. An existence theorem for bounded solutions of a.p. nonlinear elliptic equations was obtained by V.E. Slyusarchuk [77], who applied topological methods. Also, the author [62-64, 69] studied Besicovitch a.p. solutions for various classes of nonlinear partial differential equations.

Finally we must mention the existence of a.p. dynamics for nonlinear equations which are integrable by the inverse scattering method [20].

The present book deals with bounded and a.p. solutions of various evolution problems, as well as with certain stationary problems. We consider almost periodicity both in the time and space variables. In the evolution case we study first order (in the time variable) problems with monotone nonlinearities. (See, for example, [6, 27, 44, 117, 119] for monotone operators, their generalizations and applications.)

Now we will briefly sketch the contents of the book. Chapter 1 is concerned with the theory of a.p. functions. The point of view we here take is based on the notion of Bohr compactification, so our presentation differs essentially from generally accepted accounts. A similar approach is taken in [91]. Some details (especially in §§ 1-3) are taken from [91]. The principal differences are as follows. Banach space valued functions are considered (though this does not involve serious difficulties in comparison to the scalar case). Moreover, besides Bohr a.p. functions and Besicovitch a.p. functions of exponent 2, we will also consider general Besicovitch a.p. functions, Stepanov a.p. functions and weakly a.p. functions. Most results of the

Chapter are well-known, but some proofs are new. Note also that §§ 1-3 deals with functions on an arbitrary locally compact abelian group. This does not involve essential technical difficulties and is very useful for understanding relative Bohr compactifications (§ 3). Finally we introduce some spaces of smooth a.p. functions.

In Chapter 2 we state some preliminary technical results used later on. In § 1 estimates for solutions of certain integral inequalities are discussed. § 2 deals with superposition operators. Here we briefly discuss well-known facts on such operators in the Carathéodory case. However, main attention is focused on the non-Carathéodory situation both in L^p-spaces and in spaces of Besicovitch a.p. functions. Note that in the last case we make essential use of the structure of the Bohr compactification.

Chapter 3 is devoted to bounded and a.p. solution of variational inequalities. In § 1 we collect some preliminaries on such problems. The main result of § 2 is an existence theorem for bounded solutions, not involving any compactness condition. In § 3 regularity and almost periodicity are studied for the solutions we constructed. Note that almost periodicity is treated here as continuous dependence on a point of the Bohr compactification of a family of inequalities associated with the given one. Under a certain compactness condition we obtain in § 4 another existence theorem for bounded and a.p. solutions. § 5 deals with Besicovitch a.p. solutions of variational inequalities. It is important that in this case the initial formulation of the problem must be weakened in order to make sense for Besicovitch functions. In § 6 we investigate the dependence on a small parameter of bounded and a.p. solutions of singularly perturbed variational inequalities. § 7 contains a few examples. Some of these are a simple illustration of the general theory. Others are not covered by general theorems, but can be studied using additional arguments. In the same section some inequalities with retarded argument are considered as well.

Most of Chapter 4 is devoted to the investigation of evolution equations of the form

$$\frac{du}{dt}+L(t)+A(t)u = f,$$

with $L(t)$ an unbounded linear operator and $A(t)$ a monotone nonlinear operator. An abstract setting is studied in § 1 with the aid of appropriate modification of the techniques developed in § 2 and § 5 of Ch. 3. Here we do not present results based on compactness, although such results can be obtained. This is done because, as a rule, compactness conditions are not fulfilled in the most interesting applications considered in § 2. That section deals with solutions bounded and a.p. in time of positive boundary value problems for symmetric hyperbolic systems and certain nonlinear Schrödinger equations. We note that in this situation the Bohr compactification technique is essentially used. § 3 is devoted to other problems which can be studied by similar tools.

Chapter 5 deals with spatially a.p. problems. We mainly consider Besicovitch a.p. solutions. Classical solutions are very hard to study in this setting. For elliptic problems it is shown in § 1 that the monotonicity method is applicable (or not applicable) both in Sobolev spaces and in Sobolev-Besicovitch spaces. For non-linear second order elliptic equations we prove the existence of a classical bounded solution which is a.p. in the sense of Besicovitch. § 2 is devoted to (linear) systems of first order a.p. pseudodifferential operators. We prove that these are simultaneously invertible both in the L^2- and in the B^2-setting. As a consequence, B^2- solvability results are obtained for first order hyperbolic systems and single higher order hyperbolic equations. Spatially a.p. solutions of symmetric systems with monotone nonlinearity are considered in § 3 and the results of the previous section are partially extended to this case. Finally, § 4 again deals with the nonlinear Schrödinger equation.

There are also three appendices. One of them contains a list of open questions being of interest in the theory of a.p. functions and a.p. differential equations.

The bibliography is not exhaustive and mainly contains the papers cited in the text.

Now some notational remarks. The symbols $\mathbf{Z}$, $\mathbf{Q}$, $\mathbf{R}$, and $\mathbf{C}$ are used for the ring of integers and the fields of rational, real and complex numbers, respectively. We set $\mathbf{R}_+=\{x\in\mathbf{R} \mid x\geqslant 0\}$ and $\mathbf{Z}_+=\mathbf{Z}\cap\mathbf{R}_+$. The symbol $\mathbf{N}$ denotes the set of natural numbers. Standard notations are used for the usual function spaces (Lebesgue, Sobolev, etc.), multi-indices and derivatives. For spaces of E- valued functions on X we use notations of the form $\mathfrak{S}(X;E)$. If $E=\mathbf{R}$ or $\mathbf{C}$ (depending on the context) we simply use $\mathfrak{S}(X)$. According to the definition of local Lebesgue spaces we distinguish between $L^p_{\text{loc}}(0,+\infty;\cdot)$ and $L^p_{\text{loc}}(\mathbf{R}_+;\cdot)$. The latter consists of the measurable functions on $\mathbf{R}$ which belong to $L^p(0,T;\cdot)$ for all $T>0$. Notations for spaces of a.p. functions will be introduced in Ch. 1.

CHAPTER 1.

Almost periodic functions.

1. Almost periodic functions and Bohr compactification.

Recall that a topological group is a group G together with a topological space structure such that the maps $G\times G\to G$, $(x,y)\mapsto xy$ and $G\to G$, $x\mapsto x^{-1}$ are continuous. In what follows all homomorphisms of topological groups are assumed to be continuous. As a rule, only locally compact abelian groups will be considered and additive notation will be used for the group operation.

1.1. BOHR COMPACTIFICATION.

In our approach this is the most important object related with a.p. functions. The *Bohr compactification* (or *compact hull*) of the topological group G is a pair (G_B, i_B), where G_B is a compact group and $i_B\colon G\to G_B$ is a group homomorphism, such that for any homomorphism $\phi\colon G\to\Gamma$ into a compact group Γ there exist a unique homomorphism $\phi_B\colon G_B\to\Gamma$ which makes the following diagram commutative:

$$\begin{array}{ccc} G & \xrightarrow{\ \phi\ } & \Gamma \\ & {\scriptstyle i_B}\searrow \quad \swarrow{\scriptstyle \phi_B} & \\ & G_B & \end{array} \tag{1.1}$$

i.e. $\phi=\phi_B\circ i_B$. By usual categorical considerations it follows ([42]) that the Bohr compactification of a given group is determined uniquely up to isomorphism (if it exists). More precisely, if $(\tilde{G}_B, \tilde{i}_B)$ is another Bohr compactification of the group G, then there exists a unique isomorphism $j\colon G_B\to\tilde{G}_B$ such that the following diagram is commutative:

$$\begin{array}{ccc} & G & \\ {\scriptstyle i_B}\swarrow & & \searrow{\scriptstyle \tilde{i}_B} \\ G_B & \xrightarrow[\ j\ ]{} & \tilde{G}_B \end{array}$$

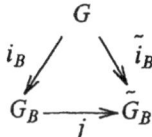

Below the group G_B itself will be called the Bohr compactification. In the sequel G denotes a locally compact abelian group. In this case the Bohr compactification can be constructed using Pontryagin's duality theory. Let G' be the *group of characters*

of G (in other words, the dual group of G). Recall that a *character* is a homomorphism from G into the multiplicative group $S^1=\{z\in\mathbf{C} \mid |z|=1\}$. The group operation in G' is given by the formula

$$(\chi_1\chi_2)(x) = \chi_1(x)\cdot\chi_2(x), \quad \chi_1, \chi_2\in G'.$$

This group is endowed with the topology of uniform convergence on compact subsets of G.

EXAMPLE (THE MAIN). $G=\mathbf{R}^n$ *is the additive group. Its characters are of the form* $\chi(x)=\exp(i\xi x)$, *where* $x\in\mathbf{R}^n, \xi\in\mathbf{R}^n, \xi x=\sum_{j=1}^n \xi_j x_j$. *The correspondence* $\xi\mapsto\exp(i\xi x)$ *defines an isomorphism* $\mathbf{R}^n\tilde{\rightarrow}(\mathbf{R}^n)'$. *In the sequel we will always identify* $(\mathbf{R}^n)'$ *with* $\mathbf{R}^n$, *using this isomorphism.*

Let $\phi: G\rightarrow H$ be a homomorphism of locally compact abelian groups. The dual homomorphism $\phi':H'\rightarrow G'$ is given by the formula $(\phi'\chi)(x)=\chi(\phi(x)), x\in G, \chi\in H'$. Therefore, passage to the group of characters defines a contravariant functor in the category of locally compact abelian groups. Set $G''=(G')'$. There is a natural map $G\rightarrow G''$, acting according to the formula $x\mapsto x(\chi)=\chi(x), \chi\in G'$. By Pontryagin's duality theorem this is an isomorphism. In the sequel the groups G and G'' are identified. Moreover, for any homomorphism ϕ the homomorphism ϕ'' coincides with ϕ.

For a topological space X the symbol X_d stands for the set X with the discrete topology. The identity map $X_d\rightarrow X$ is continuous (but is not a homeomorphism in general).

THEOREM 1.1. *Let* $G_B=(G')'_d$ *and let* $i_B: G=G''\rightarrow G_B$ *be the homomorphism dual to the identity* $(G')_d\rightarrow G'$. *Then the pair* (G_B, i_B) *is the Bohr compactification of G.*

Proof. Let us show that $i_B(G)$ is dense in G_B.

Let $\chi\in G_B$ and $\chi|_{i_B(G)} = 1$. Then $(i_B)'(\chi)=1$. By the duality theorem, $(i_B)'$ is the identity map from $G_B=(G')_d$ to G' and therefore $\chi\equiv 1$. If now $\overline{i_B(G)}\neq G_B$, then any nontrivial character of the group $G_B/i_B(B)$ lifts to some nontrivial character of the group G_B which equals 1 on $i_B(G)$. This contradiction completes the proof of denseness.

Now we check the universality property. It is sufficient to establish the existence of the homomorphism ϕ_B in (1.1); its uniqueness follows from the denseness of the image of i_B. Without loss of generality we may assume that the group Γ is abelian, otherwise Γ may be replaced by $\overline{\phi(G)}$. Consider a homomorphism $\phi:\Gamma'\rightarrow G'$. The group Γ' is discrete since Γ is compact. Therefore ϕ' may be regarded as a homomorphism $\phi':\Gamma'\rightarrow(G')_d$. The required homomorphism $\phi_B:(G')'_B=G_B\rightarrow\Gamma''=\Gamma$ is obtained by duality. The commutativity of (1.1) is obvious. □

We must point out that the homomorphism i_B is injective. This follows from the surjectivity of the identity map $(G')_d\rightarrow G'$. Further, we will often identify the elements of G with their images in G_B.

1.2. UNIFORMLY ALMOST PERIODIC FUNCTIONS.

Let E be a Banach space (real or complex). Denote by $C(G;E)$ the space of all continuous functions from G to E, and by $C_b(G;E)\subset C(G;E)$ the subspace of bounded functions. The space $C(G;E)$ is endowed with the topology of uniform convergence on compact sets, and $C_b(G;E)$ is equipped with the norm

$$||f||_{C_b}=\sup_{x\in G}||f(x)||_E, \tag{1.2}$$

which gives it a Banach space structure.

A subset $T\subset G$ is said to be *relatively dense* if there is a compactum $K\subset G$ such that $(S+K)\bigcap T=\varnothing$ for all $s\in G$. A function $f\in C_b(G;E)$ is said to be *Bohr almost periodic* or *uniformly almost periodic* if for any $\epsilon>0$ the set

$$T_f(\epsilon)=\{\tau\in G\,|\,\sup_{x\in G}||f(x+\tau)-f(x)||\leqslant\epsilon\}$$

of *ϵ-almost periods* of f is relatively dense in G. The Bohr a.p. functions constitute the subspace $\mathrm{CAP}(G;E)\subset C_b(G;E)$. In the case $E=\mathbf{R}$ or $\mathbf{C}$ we shall simply write $\mathrm{CAP}(G)$.

The following assertion is the main result of the theory of a.p. functions.

THEOREM 1.2. *For $f\in C_b(G;E)$ the following three statements are equivalent:*

a) $f\in\mathrm{CAP}(G;E)$;

b) the set of translates $\{f(\cdot+\tau)\,|\,\tau\in G\}$ is relatively compact in $C_b(G;E)$;

c) there exists a function $\bar{f}\in C(G_B;E)$ such that $f=i_B^*\bar{f}=\bar{f}\circ i_B$, i.e. f can be extended to a continuous function on G_B.

Proof. The group G_B acts on continuously on $C(G_B;E)$ by translation of arguments. Since G_B is compact, the implication c) $\Rightarrow$ b) follows immediately.

b) $\Rightarrow$ c). Consider the metric compactum $H(f)=\overline{\{f(\cdot+\tau)\}}\subset C_b(G;E)$. Obviously, the action of the group G on $C_b(G;E)$ by translation of arguments is isometric. The compactum $H(f)$ is invariant under this action. Therefore, there is a well-defined homomorphism from G to the group Γ of isometries of $H(f)$. The group is compact in the topology of uniform convergence by the Arzelà-Ascoli theorem. By the universality of G_B there exists an extension $G_B\to\Gamma$ of the homomorphism $G\to\Gamma$. In other words, the function f may be shifted by the elements of G_B and then $f(0+\tau),\tau\in G_B$, gives the required extension $\bar{f}$ of f to G_B.

a) $\Rightarrow$ b). The function f is uniformly continuous. Indeed, for any $\epsilon>0$ there is a compactum $K\subset G$ such that $(s+K)\bigcap T_f(\epsilon/3)\neq\varnothing$ for all $s\in G$. Take a relatively compact symmetric (i.e. $V=-V$) neighbourhood V of zero in G. The function f is uniformly continuous on the compactum $K+K+\bar{V}$. The neighbourhood V may be selected in such a way that

$$||f(x)-f(y)||_E \leqslant \epsilon/3, \quad x,y\in K+K+\overline{V},\ x-y\in V.$$

Then for any $x,y\in G, x-y\in V$ we have the inequality $||f(x)-f(y)||_E\leqslant\epsilon$. Indeed, let $x\in s+K$ and $\tau\in(-s+K)\cap T_f(\epsilon/3)$. As $x+\tau\in K+K$, $x-y=(x+\tau)-(y+\tau)\in V$ and $y+\tau=x+\tau-(x-y)\in K+K+\overline{V}$, we have

$$||f(x)-f(y)||_E \leqslant ||f(x)-f(x+\tau)||_E+||f(x+\tau)-f(y+\tau)||_E+$$

$$+||f(y+\tau)-f(y)||_E \leqslant \epsilon$$

This proves the uniform continuity of f.

In order to prove the relative compactness of the set of translates we will construct for arbitrary $\epsilon>0$ a finite ϵ-net. Take a compact set $K\subset G$ and a symmetric neighbourhood of zero $U\subset G$ such that $(s+K)\cap T_f(\epsilon/2)\neq\varnothing$ for all $s\in G$ and $||f(x)-f(y)||\leqslant\epsilon/2$ for $x-y\in U$. Choose a finite covering of K by sets $\tau_j+U, j=1,\ldots,N$. For an arbitrary $\tau\in G$ there is a $\tau'\in(-\tau+K)\cap T_f(\epsilon/2)$, and so $||f(\cdot+\tau)-f(\cdot+\tau'+\tau)||_{C_b}\leqslant\epsilon/2$. But $\tau'+\tau\in K$, and, consequently, there is a τ_j such that $\tau'+\tau-\tau_j\in U$. Hence it follows that $||f(\cdot+\tau'+\tau)-f(\cdot+\tau_j)||_{C_b}\leqslant\epsilon/2$. Then $||f(\cdot+\tau)-f(\cdot+\tau_j)||C_b\leqslant\epsilon$ and the translates $\{f(\cdot+\tau_j)\}_{j=1}^N$ turns out to form the required ϵ-net.

b) $\Rightarrow$ a). Let $\{f(\cdot+\tau_j)\}_{j=1}^N$ be a finite ϵ-net in the set of translates. Then for any $\tau\in G$ we can find a τ_j such that $||f(\cdot+\tau)-f(\cdot+\tau-\tau_j)||_{C_b}\leqslant\epsilon$, or, which is the same, $||f(\cdot)-f(\cdot+(\tau-\tau_j))||_{C_b}\leqslant\epsilon$. Hence the set $\{\tau-\tau_j\}$ contains an ϵ-almost period. In other words, any translate of the compactum $K=\{\tau_1,\ldots,\tau_N\}$ has nonempty intersection with $T_f(\epsilon)$, that is, $T_f(\epsilon)$ is relatively dense. □

The equivalence of a) and b) is known as Bochner's criterion of almost periodicity. Note that the extension $\tilde{f}$ is defined uniquely, since $i_B(G)$ is dense in G_B. For the same reason the following equality holds:

$$\sup_{x\in G}||f(x)||_E = \sup_{y\in G_B}||\tilde{f}(y)||_E. \tag{1.3}$$

Thus the map $i_B^*:\tilde{f}\mapsto\tilde{f}\circ i_B$ is an isometric isomorphism

$$i_B^*:C(G_B;E)\overset{\sim}{\rightarrow}\mathrm{CAP}(G;E). \tag{1.4}$$

In the sequel these spaces are often identified.

The following properties of a.p. functions follow from the identification (1.4) and corresponding properties of spaces of continuous functions.

1. CAP$(G;E)$ *is a Banach space with the norm (1.2).*

2. CAP(G) *is a commutative Banach algebra having* G_B *as maximal ideal space. For an arbitrary space E the space* CAP$(G;E)$ *is a Banach* CAP(G)*-module*[*]

3. For a bounded linear operator $A:E\to F$ *the formula* $(A_*f)(x)=Af(x)$, $x\in G$, $f\in$CAP$(G;E)$, *defines a homomorphism* A_*:CAP$(G;E)\to$CAP$(G;E)$ *of Banach* CAP(G)*-modules such that* $||A_*||=||A||$.

4. If $g:E\to F$ *is a continuous (not necessarily linear) transformation, then* $g\circ f\in$CAP$(G;F)$ *for* $f\in$CAP$(G;E)$.

We denote by Trig$(G;E)$ the space of "trigonometric polynomials" with coefficients in E: it is the subspace of CAP$(G;E)$ consisting of the functions $v_1\chi_1+\cdots+v_N\chi_N$, where $\chi_j\in G', v_j\in E; j=1,\ldots,N$; and N is any natural number. In the case $G=\mathbf{R}^n$ these are the usual trigonometric polynomials. The following assertion is known as the approximation theorem.

PROPOSITION 1.3. *The space* Trig$(G;E)$ *is dense in* CAP$(G;E)$.

Proof. All characters of the group G can be obtained by restriction of the characters of the group G_B. Applying the Peter-Weyl theorem [87] to the compact group G_B, we find that linear combinations of characters are dense in $C(G_B)$. Together with the identification (1.4) this gives the required statement for the case $E=\mathbf{C}$ (or $\mathbf{R}$).

In general, the isomorphism (1.4) together with the well- known isomorphism $C(X;E)\simeq C(X)\otimes_\epsilon E$, where X is compact and $\otimes_\epsilon$ is the complete topological tensor ϵ-product ([83, 131]), implies that CAP$(G;E)\simeq$ CAP$(G)\otimes_\epsilon E$. It is easy to see that the restriction of this isomorphism to Trig$(G;E)$ gives the isomorphism Trig$(G;E)\simeq$ Trig$(G)\otimes E$ (in the algebraic sense). The required statement now follows from the case which has been considered above and from the continuity of the canonical map $E\times F\to E\otimes_\epsilon F$, $(u,v)\mapsto u\otimes v$. □

Below a stronger result will be presented (Theorem 2.4). In its proof only the scalar version of the previous result will be used.

REMARK 1.4. The definition of Bohr almost periodicity generalises immediately to maps $f: G\to Y$ where Y is a complete metric space. Moreover, a statement similar to Theorem 1.4 is valid (with essentially the same proof). In particular a map $f: G\to Y$ is a.p. if and only if (iff) it may be extended to a continuous map $f:G_B\to Y$. Obviously, in this case the set $f(G)$ is relatively compact in Y.

[*]Recall that a Banach C-module (with C a commutative Banach algebra) is a Banach space F with a bilinear and associative multiplication $C\times F\to F$, $(c,f)\to c\cdot f$ for which $||c\cdot f||\leqslant||c||\cdot||f||$. Homomorphisms of C-modules are exactly the C-linear maps.

2. Besicovitch almost periodic functions and harmonic analysis.

2.1. MEAN VALUE.

Let μ be the normalized Haar measure on G_B, that is, the positive regular Borel measure such that $\mu(U)=\mu(U+s)$ for every Borel subset $U\subset G_B$ and any $s\in G_B$ (invariance) and $\mu(G_B)=1$ (normalization). Since G_B is a compact abelian group, such a measure exists and is unique ([87]).

The *mean value* $\mathbf{M}\{f\}$ of a function $f\in\mathrm{CAP}(G;E)$ is defined by the formula

$$M\{f\} = \int_{G_B} \tilde{f}(x)d\mu(x), \tag{2.1}$$

where $\tilde{f}\in C(G_B;E)$ is the extension of f to G_B. The following properties of the mean value are consequences of standard properties of the Haar measure and of the integral.

1. $\mathbf{M}:\mathrm{CAP}(G;E)\to E$ is a bounded linear operator, and, moreover, $||\mathbf{M}\{f\}||_E\leqslant||f||_{\mathrm{CAP}}$.

2. For arbitrary $f\in\mathrm{CAP}(G;E)$ and $s\in G$ the equality $\mathbf{M}\{f(\cdot+s)\}=\mathbf{M}\{f\}$ holds (invariance).

3. If $u\in E$ and $f\in\mathrm{CAP}(G)$, then $\mathbf{M}\{f\cdot u\}=\mathbf{M}\{f\}\cdot u$.

4. $\mathbf{M}\{1\}=1$.

5. If $f\in\mathrm{CAP}(G)$ and $f>0$, then $\mathbf{M}\{f\}\geqslant 0$.

6. If $A:E\to F$ is a bounded linear operator, then

$$\mathbf{M}\{A\cdot f\} = A\mathbf{M}\{f\}.$$

PROPOSITION 2.1. *Let $f\in\mathrm{CAP}(G)$. If $f>0$ and $\mathbf{M}\{f\}=0$, then $f\equiv 0$.*

Proof. The extension $\tilde{f}\in C(G_B)$ of the function f is clearly nonnegative. If f, and so $\tilde{f}$, is nonzero, then for some open $U\subset G_B$ the inequality $\tilde{f}(x)\geqslant\alpha>1$ holds for $x\in U$. Then $\mathbf{M}\{f\}\geqslant\alpha\mu(U)>0$. The contradiction obtained completes the proof. □

The case $G=\mathbf{R}^n$. Put $K_T=\{x\in\mathbf{R}^n \mid |x_j|\leqslant T, j=1,\ldots,n\}$.

THEOREM 2.2. *For $f\in\mathrm{CAP}(\mathbf{R}^n;E)$,*

$$\mathbf{M}\{f\} = \lim_{T\to\infty}\frac{1}{(2T)^n}\int_{s+K_T} f(x)dx, \tag{2.2}$$

where the limit exists uniformly in $s\in\mathbf{R}^n$.

Proof. By property 1 and Proposition 1.3 it suffices to check equality (2.2) and the fact that the limit is uniform on characters, which are in this case exponentials $\exp(i\xi\cdot x)$. In this case we have

$$\int_{s+K_T} \exp(i\xi\cdot x)dx = \exp(i\xi\cdot s)\int_{K_T}\exp(i\xi\cdot x)dx$$

and a straightforward computation shows that this integral equals $(2T)^n$ if $\xi=0$, while its modulus can be estimated by $(2T)^{n-k}$ if $\xi\neq 0$ (here k is the number of nonzero components of the vector ξ). This gives (2.2). □

The representation (2.2) has some useful versions. For example, let $\{\Omega_k \mid k\in\mathbf{Z}_+\}$ be a sequence of bounded regions in $\mathbf{R}^n$. For a region $\Omega\subset\mathbf{R}^n$ we denote by $\text{mes}\,\Omega$ its Lebesgue measure and we put

$$(\Omega)_h = \{x\in\Omega \mid \text{the distance of } x \text{ to } \partial\Omega \text{ is} \leqslant h\}.$$

Assume that $\text{mes}\,\Omega_k\to\infty$ and that there exist $h_k\in\mathbf{R}$, $h_k\to\infty$, such that

$$\lim\frac{\text{mes}(\Omega_k)_{h_k}}{\text{mes}\,\Omega_k} = 0.$$

Then

$$\mathbf{M}\{f\} = \lim\frac{1}{\text{mes}\,\Omega_k}\int_{\Omega_k} f(x)dx. \tag{2.3}$$

In particular, if $V_R=\{x\in\mathbf{R}^n \mid |x|\leqslant R\}$ is a ball of radius R, then

$$\mathbf{M}\{f\} = \lim_{R\to\infty}\frac{1}{\text{mes}V_R}\int_{s+V_R} f(x)dx \tag{2.4}$$

uniformly in $s\in\mathbf{R}^n$. For the proof see [89].

2.2. BESICOVITCH ALMOST PERIODIC FUNCTIONS.

For many reasons it is useful to extend the class of almost periodic functions in the following way. By $B^p(G;E)$, $1\leqslant p\leqslant\infty$, we denote *the space of Besicovitch a.p. functions* with values in a Banach space E, that is, the space $L^p(G_B;E)$, where L^p is taken with respect to the Haar measure on G_B. The norm in $B^p(G;E)$ is given by the formula

$$\|f\|^p_{B^p} = \mathbf{M}\{\|f\|^p_E\} = \int_{G_B}\|f(x)\|^p_E d\mu(x) \tag{2.5}$$

for $1\leqslant p<\infty$. As usual, if $p=\infty$,

$$\|f\|_{B^\infty} = \operatorname*{ess\,sup}_{x\in G_B}\|f(x)\|_E.$$

For Besicovitch a.p. functions the definition of mean value (2.1) makes sense too.

The following properties of Besicovitch a.p. functions are obvious specializations of properties of L^p spaces [95, Ch. 8].

1. $B^p(G;E)$ *is a Banach space with norm (2.5). The identification of* $\mathrm{CAP}(G;E)$ *with* $C(G_B;E)$ *defines a continuous embedding* $\mathrm{CAP}(G;E) \subset B^p(G;E)$ *which is dense if* $p<\infty$. *Moreover,* $||f||_{B^p} \leq ||f||_{\mathrm{CAP}}$.

2. If E is reflexive, then the dual of $B^p(G;E)$, $1 \leq p < \infty$, *is the space* $B^p(G;E)' = B^{p'}(G;E)$, *where* $1/p+1/p'=1$. *The pairing is given by the formula*

$$(u,v)_B = \mathbf{M}\{(u,v)\} = \int_{G_B} (u(x), v(x))dx, \tag{2.6}$$

where $(\cdot,\cdot)$ *is the pairing between* E' *and* E. *It especially follows that for* $p>1$ *and reflexive E the space* $B^p(G;E)$ *is reflexive - a fact which will be very useful in the future. In the case when* $E=H$ *is a Hilbert space, the space* $B^2(G;H)$ *is also a Hilbert space, with scalar product defined by (2.6).*

3. $B^p(G;B)$ *is a Banach* $\mathrm{CAP}(G)$*- module with respect to the natural operations.*

4. For a bounded linear operator $A:E \to F$ *the formula* $(A_* u)(x) = Au(x)$ *defines a homomorphism of Banach* $\mathrm{CAP}(G)$*-modules* $A_*: B^p(G;E) \to B^p(G;F)$ *such that* $||A_*|| = ||A||$.

5. There is a continuous and dense embedding $B^r(G;E) \subset B^p(G;E)$ *for* $p<r$.

6. $\mathbf{M}: B^p(G;E) \to E$ *is a bounded linear operator such that* $||\mathbf{M}\{f\}||_E \leq ||f||_{B^p}$.

The case $G=\mathbf{R}^n$. For $p \neq \infty$ we give another description of the space $B^p(\mathbf{R}^n;E)$ (in essence it coincides with the classical definition [40]; for the definition of Besicovitch a.p. functions in terms of ϵ-almost periods see [106]). For $f \in L^1_{\mathrm{loc}}(\mathbf{R}^n)$, $f>0$ we put

$$\overline{\mathbf{M}}\{f\} = \limsup_{T\to\infty} \frac{1}{(2T)^n} \int_{K_T} f(x)dx \tag{2.7}$$

(K_T has been defined above Theorem 2.1). We denote by $LM^p(\mathbf{R}^n;E)$, $1 \leq p < \infty$, the subspaces of $L^p_{\mathrm{loc}}(\mathbf{R}^n;E)$ consisting of functions that have the finite seminorm

$$<f>_p = \overline{\mathbf{M}}\{||f||_E^p\}^{1/p}, \tag{2.8}$$

and by $LM^p_0(\mathbf{R}^n;E)$ the subspace of functions that have seminorm (2.8) zero, that is, the closure of zero in $LM^p(\mathbf{R}^n;E)$. In the quotient

$$\mathcal{B}^p(\mathbf{R}^n;E) = LM^p(\mathbf{R}^n;E)/LM^p_0(\mathbf{R}^n;E)$$

the seminorm (2.8) induces a norm.

PROPOSITION 2.3. *$\mathcal{B}^p(\mathbf{R}^n; E)$ is a Banach space.*

Proof. The case $n=1$ of this statement is in essence contained in Theorem 5.10.1 of [40], Part 1. The general case is quite similar to the case $n=1$. □

Note that $\mathrm{CAP}(\mathbf{R}^n; E) \cap LM_0^p(\mathbf{R}^n; E)=\{0\}$ (by property 5, n°. 2.1). Hence composition of the standard embedding $\mathrm{CAP}(\mathbf{R}^n; E) \subset LM^p(\mathbf{R}^n; E)$ and the quotient map gives a continuous embedding $\mathrm{CAP}(\mathbf{R}^n; E) \subset \mathcal{B}^p(\mathbf{R}^n; E)$. Now $B^p(\mathbf{R}^n; E)$ may be defined as the closure of $\mathrm{CAP}(\mathbf{R}^n; E)$ in $\mathcal{B}^p(\mathbf{R}^n; E)$. The equivalence of these two definitions is evident (using property 1 of n°. 2.2), since $||\cdot||_p$ and $<\cdot>$ coincide on $\mathrm{CAP}(\mathbf{R}^n; E)$.

2.3. BOCHNER-FEJÉR OPERATORS.

By the approximation theorem (Proposition 1.3) and assertion 1 of n°. 2.2, the space $\mathrm{Trig}(G; E)$ is dense in each of the spaces $\mathrm{CAP}(G; E)$ and $B^p(G; E)$, $1 \leqslant p < \infty$. We will strengthen this result using linear operators which realize approximations of a.p. functions by trigonometric polynomials (so-called *Bochner-Fejér operators*).

We construct a δ-shaped net of functions on the Bohr compactification G_B. Denote by $\mathfrak{U}$ the inclusion-directed set consisting of all neighbourhoods of zero of the group G_B. According to Urysohn's lemma, for any neighbourhood $U \in \mathfrak{U}$ there exists a function $\phi_U \in C(G_B) = \mathrm{CAP}(G_B)$ such that $\phi_U \geqslant 0$, $\mathrm{supp}\, \phi_U \subset U$ and $\mathbf{M}\{\phi_U\}=1$. By Proposition 1.3, for any $U \in \mathfrak{U}$ and $\epsilon>0$ there exists a function $\psi_{U,\epsilon} \in \mathrm{Trig}(g)$ such that

$$\sup_{x \in G} | \phi_U(x) - \psi_{U,\epsilon}(x) | < \epsilon.$$

Note that here we have used only the scalar version of Proposition 1.3. We put

$$\lambda_{U,\epsilon} = (\mathbf{M}\{\psi_{U,\epsilon}\}+\epsilon)^{-1}(\psi_{U,\epsilon}+\epsilon) \in \mathrm{Trig}(G)$$

and define the directed set $\Gamma = \mathfrak{U} \times (0,1)$ with order $\prec$, where $\gamma_1 = (U_1, \epsilon_1) \prec \gamma_2 = (\mathfrak{U}_2, \epsilon_2)$ iff $U_1 \supset U_2$, $\epsilon_2 \leqslant \epsilon_1$. The net of trigonometric polynomials $\{\lambda_\gamma\}$ possesses the following properties, which mean that it is δ-shaped:

a) $\lambda_\gamma(x) \geqslant 0$, $x \in G_B$;

b) $\mathbf{M}\{\lambda_\gamma\}=1$;

c) $\lim_\Gamma(\sup_{x \in G_B \setminus V} \lambda_\gamma(x))=0$ for any neighbourhood V of zero in the group G_B.

For an a.p. function $f \in B^1(G; E)$ we put

$$(L_\gamma f)(x) = \mathbf{M}_y\{\lambda_\gamma(y) f(x-y)\} = \mathbf{M}_y\{\lambda_\gamma(x-y) f(y)\}. \tag{2.9}$$

In other words,

$$L_\gamma f = \lambda_\gamma \star f, \tag{2.9'}$$

where the convolution is taken in the group G_B. Expression (2.9) together with properties a) and b) of the kernels $\{\lambda_\gamma\}$ prove the estimate

$$|| L_\gamma f ||_{\mathrm{CAP}} \leqslant || f ||_{\mathrm{CAP}} f \in \mathrm{CAP}(G; E). \tag{2.10}$$

The Hausdorff-Young inequality

$$|| h \star g ||_{B^p} \leqslant || h ||_{B^1} || g ||_{B^p}$$

(which holds on an arbitrary locally compact abelian group), together with properties a) and b), implies the estimate

$$|| L_\gamma f ||_{B^p} \leqslant || f ||_{B^p}, \quad f \in B^p(G; E), \quad 1 \leqslant p < \infty. \tag{2.11}$$

Thus, the net $\{L_\gamma\}$ of linear operators acting in the spaces $\mathrm{CAP}(G; E)$ and $B^p(G; E)$, $1 \leqslant p < \infty$, is uniformly bounded in $\gamma \in \Gamma$.

We put

$$\lambda_\gamma(x) = \sum_{\chi \in (G')_d} c_{\gamma,\chi} \cdot \chi(x)$$

(the sum is finite). The second equality in (2.9) implies

$$(L_\gamma f)(x) = \mathbf{M}_y \{[\sum c_{\gamma,\chi} \cdot \chi(x) \chi(-y)] f(y)\} =$$

$$= \sum c_{\gamma,\chi} \cdot \mathbf{M}\{\overline{\chi}(y) \cdot f(y)\} \cdot \chi(x). \tag{2.12}$$

Thus,

$$L_\gamma f \in \mathrm{Trig}(G; E). \tag{2.13}$$

Observe that the operators L_γ are finite dimensional in the case $E = \mathbf{C}$. It is also easy to see that the L_γ commute with bounded operators acting on the ranges of the functions we consider: for $A \in L(E, F)$,

$$L_\gamma A = A L_\gamma.$$

The following important result is valid.

THEOREM 2.4. *Let* $\mathfrak{F} \subset \mathrm{CAP}(G; E)$ *(respectively,* $\mathfrak{F} \subset B^p(G; E)$, $1 \leqslant p < \infty$*) be a relatively compact subset. Then* $\lim_\Gamma L_\gamma f = f$, *uniformly in* $f \in \mathfrak{F}$, *in the norm of the* *corresponding space.*

Proof. Let $f \in \mathrm{CAP}(G; E)$. Due to property b) of the kernels $\{\lambda_\gamma\}$ we have

$$f(x)-(L_\gamma f)(x) = \mathbf{M}_y\{\lambda_\gamma(y)[f(x)-f(x-y)]\} =$$

$$= \int_V \lambda_\gamma(y)[f(x)-f(x-y)]d\mu(y) + \int_{G_B\setminus V} \lambda_\gamma(y)[f(x)-f(x-y)]d\mu(y) =$$

$$= \mathcal{G}_1(x)+\mathcal{G}_2(x).$$

Here f is extended to a function belonging to $C(G_B; E)$, and V is a neighbourhood of zero in G_B. By the uniform continuity of f on G_B we have

$$\| f(x)-f(x-y) \|_E \leqslant \epsilon, \quad y \in V, x \in G_B,$$

for a sufficiently small neighbourhood V. Then

$$\| \mathcal{G}_1(x) \|_E \leqslant \epsilon \int_V \lambda_\gamma(y)d\mu(y) \leqslant \epsilon \cdot \mathbf{M}\{\lambda_\gamma\} = \epsilon$$

and

$$\| \mathcal{G}_2(x) \|_E \leqslant 2 \| f \|_{\mathrm{CAP}} \cdot \sup_{x \in G_B\setminus V} \lambda_\gamma(x).$$

The right-hand side of the last inequality tends to zero by property b) of the kernels $\{\lambda_\gamma\}$. Hence it follows that $L_\gamma f \to f$ in $\mathrm{CAP}(G; E)$, because ϵ has been taken arbitrarily.

We proceed to establish that the limit is uniform in $f \in \mathcal{F} \subset \mathrm{CAP}(G; E)$. Choose a finite ϵ-net $\{f_j\} \subset \mathcal{F}$, $j=1, \ldots, N$. Then by (2.10) we have

$$\| f-L_\gamma f \|_{\mathrm{CAP}} \leqslant$$

$$\leqslant \| f-f_j \|_{\mathrm{CAP}} \| f_j-L_\gamma f_j \|_{\mathrm{CAP}} + \| L_\gamma(f-f_j) \|_{\mathrm{CAP}} \leqslant$$

$$\leqslant 2\epsilon + \| f_j-L_\gamma f_j \|_{\mathrm{CAP}}$$

for some f_j. As we have proved above, the second summand at the right-hand side of this inequality tends to zero for any $j=1, \ldots, N$. Because ϵ has been taken arbitrarily, this concludes the proof.

The case $\mathfrak{F}\subset B^p(G;E)$, $1\leqslant p<\infty$, is quite similar. It is necessary to find a finite ϵ-net $\{f_j\}_{j=1}^N$ which belongs to CAP$(G;E)$. This can be done using assertion 1 of n°. 2.2. By using (2.11) we can obtain, as before, the following estimate:

$$||f-L_\gamma f||_{B^p} \leqslant 2\epsilon+||f_j-L_\gamma f_j||_{CAP}$$

for some $j=1,\ldots,N$. This gives the required statement. □

REMARK 2.5. We give a modified construction of the operators L_γ, which is very useful for technical reasons. Let $\mathfrak{U}^{(s)}\subset\mathfrak{U}$ be a directed subset consisting of symmetric neighbourhoods of zero and let $\Gamma^{(s)}$ be the corresponding subset of Γ. It is easy to check that for $\gamma=(\mathfrak{U},\epsilon)\in\Gamma^{(s)}$ the functions ϕ_U,ψ_γ and λ_γ defined above may taken even: $\phi_U(x)=\phi_U(-x)$, $\psi_\gamma(x)=\psi_\gamma(-x)$, $\lambda_\gamma(x)=\lambda_\gamma(-x)$. Properties a) - b), and consequently Theorem 2.4, still hold with Γ replaced by $\Gamma^{(s)}$. Below we will omit the symbol s and assume that $\mathfrak{U}$ consists of symmetric neighbourhoods and that the functions λ_γ are even if necessary. In this situation the Bochner-Fejér operators are symmetric relatively to the scalar product $(\cdot,\cdot)$ defined in (2.6):

$$(L_\gamma g,f)_B = (g,L_\gamma f)_B,\ f\in B^p(G;E),\ g\in B^{p'}(G;E). \tag{2.14}$$

2.4. FOURIER-BOHR TRANSFORM.

In essence, harmonic analysis of a.p. functions reduces to harmonic analysis on the Bohr compactification. The *Fourier-Bohr transform* F_B is given by the formula

$$(F_Bf)(\chi) = \mathbf{M}\{f\cdot\overline{\chi}\},\ \ \chi\in G_B' = (G')_d. \tag{2.15}$$

This is simply the Fourier transform on the group G_B, and sometimes the notation $\hat{f}=F_Bf$ will be used. Formula (2.15) make sense if, for example, $f\in B^1(G;E)$.

The set

$$\mathrm{sp}(f) = \{\chi\in(G')_d\ |\hat{f}(\chi)\neq 0\}$$

is called *the spectrum of the a.p. function f*. The subgroup $S(f)\subset(G')_d$ generated by the set sp(f) is called the *frequency group of the a.p. function f*. The following important inclusion derives from (2.12):

$$\mathrm{sp}(L_\gamma f)\subset\mathrm{sp}(f). \tag{2.16}$$

Consequently,

$$S(L_\gamma f) \subset S(f). \tag{2.17}$$

PROPOSITION 2.6. *For $f \in B^1(G;E)$ the sets* sp(f) *and* $S(f)$ *are at most countable.*

Proof. It suffices to show that sp(f) is at most countable. Due to Proposition 1.3 and assertion 1 of n°. 2.2, Trig($G;E$) is dense in $B^1(G;E)$. Let us approximate f by elements $f_k \in$Trig($G;E$) in the B^1-norm. The set sp(f_k) is finite. Indeed, it is the set of all $\chi \in (G')_d$ which appear in f_k with nonzero coefficient. By the continuity of **M** on $B^1(G;E)$, the inclusion sp(f)$\subset \bigcup$ sp(f_k) holds. This gives the statement required. □

Let I be an arbitrary index set. We denote by $m(I;E)$ the Banach space of all bounded functions $\phi: I \to E$ with norm

$$||\phi||_m = \sup_{i \in I} ||\phi(i)||_E.$$

This space contains the closed subspace $l^\infty(I,E)$ consisting of the functions equal to zero outside some countable subset of I (depending on the function). Also, we denote by $l^p(I;E)$, $1 \leqslant p < \infty$, the Banach space of functions $\phi: I \to E$ having finite norm

$$||\phi||_{l^p} = (\sum_{i \in I} ||\phi(i)||_E^p)^{1/p}.$$

Obviously, $l^p(I;E) \subset l^q(I;E)$ if $q \geqslant p$. This inclusion is continuous, and it is dense if $q < \infty$.

We return to the Fourier-Bohr transform. Let $f \in B^1(G;E)$. Then for any character $\chi \in (G')_d$,

$$||(F_B f)(\chi)||_E \leqslant \mathbf{M}\{||f||_E\} = ||f||_{B^1}. \tag{2.18}$$

Furthermore, $F_B f$ vanishes outside the countable set sp(f). Therefore the Fourier-Bohr transform is a bounded linear operator

$$F_B : B^1(G;E) \to l^\infty((G')_d;E).$$

Its norm equals one. Indeed, (2.18) implies that $||F_B|| \leqslant 1$. But for $f \equiv 1$ we have $(F_B f)(\chi) = 1$ if $\chi = 1$ and $(F_B f)(\chi) = 0$ if $\chi \neq 1$. Thus $||F_B|| = 1$.

The following statement is known as the uniqueness theorem for a.p. functions.

PROPOSITION 2.7. *The kernel of F_B is zero.*

Proof. Let $f \in \ker F_B$. Then sp(f)$= \varnothing$. Consider the trigonometric polynomials $f_\gamma = L_\gamma f$. By (2.16) we have sp(f_γ)$= \varnothing$ for all $\gamma \in \Gamma$. This trivially implies that

$f_\gamma \equiv 0$ for any $\gamma \in \Gamma$. Now the statement required follows from Theorem 2.4. □

In the general case an exact description of the space $F_B(B^p(G;E)$ is unknown. But the orthogonality of characters and the Peter-Weyl theorem gives

PROPOSITION 2.8. *If H is a Hilbert space, then $F_B:B^2(G;H) \to l^2((G')_d;H)$ is an isometric isomorphism.* □

3. Structure of Bohr compactifications and spaces of almost periodic functions.

3.1. REPLACEMENT OF THE BASIC GROUP AND THE KRONECKER- WEYL THEOREM.

Here we investigate the behaviour of the Bohr compactification under replacement of G. Let $\phi:G \to H$ be a homomorphism of locally compact abelian groups. The characteristic property of the Bohr compactification, when applied to the composition $G \to H \to H_B$, implies the existence and uniqueness of a homomorphism $\tilde{\phi}:G_B \to H_B$ such that the diagram

$$\begin{array}{ccc} G & \xrightarrow{\ \phi\ } & H \\ {\scriptstyle i_B}\downarrow & & \downarrow{\scriptstyle i_B} \\ G_B & \xrightarrow[\ \tilde{\phi}\ ]{} & H_B \end{array} \qquad (3.1)$$

commutes. Hence the map

$$\phi^* = \phi_E^*:\mathrm{CAP}(H;E) \to \mathrm{CAP}(G;E)$$

given by the formula $\phi^* f = f \circ \phi$ is well- defined. Indeed, by the commutativity of the diagram (3.1) we have: $f \circ \phi = (\tilde{f} \circ \tilde{\phi})\,|_{i_B(G)}$, where $\tilde{f}$ is the canonical extension of f to H_B. It is easy to verify that ϕ^* is a linear operator, with norm one.

Due to the almost periodicity of the characters of H we have $H' \subset \mathrm{CAP}(H)$. The restriction of ϕ^* to H' coincides with the Pontryagin duality $\phi':H' \to G'$.

The following theorem is essentially a general form of the Kronecker-Weyl theorem.

THEOREM 3.1. *The following assertions are equivalent:*

a) $\overline{\phi(G)} = H$;

b) if ϕ' is the homomorphism of dual groups, then the kernel $\ker \phi'$ *is trivial:*

c') for some Banach space E the operator ϕ_E^* *is injective:* $\ker\phi_E^*=\{0\}$*;*

c) $\ker\phi_E^*=\{0\}$ *for any Banach space E;*

d') for some Banach space E the equality

$$\mathbf{M}\{f\} = \mathbf{M}\{\phi_E^* f\}, \quad f\in \mathrm{CAP}(H;E), \tag{3.2}$$

holds;

d) for any Banach space E the equality (3.2) holds.

Proof. The implications c) $\Rightarrow$ c') and d) $\Rightarrow$ d') are trivial, while the equivalence a) $\Leftrightarrow$ b) is an easy consequence of Pontryagin's duality theorem.

c') $\Rightarrow \ker\phi_{\mathbf{C}}^*=\{0\}$. Let $u_0\in E, u_0\neq 0$. Consider the linear embedding $\mathrm{CAP}(\cdot)\to\mathrm{CAP}(\cdot, E), f\mapsto f\cdot u_0$. The map ϕ^* commutes with this embedding, hence $\ker\phi_{\mathbf{C}}^*=\{0\}$.

c') $\Rightarrow$ b). This follows from $\ker\phi_{\mathbf{C}}^*=\{0\}$ and $\phi'=\phi_{\mathbf{C}}^*|_{H'}$.

c') $\Rightarrow$ c). For any $v\in E'$ we have $\phi_{\mathbf{C}}^*(v,f)=(v,\phi_E^* f)$. Therefore $(v,f)\in\ker\phi_{\mathbf{C}}^*$ if $f\in\ker\phi_E^*$. In our case $\ker\phi_{\mathbf{C}}^*=\{0\}$, and this implies the assertion desired, because v has been taken arbitrarily.

d') $\Rightarrow$ c'). Let $f\in\mathrm{CAP}(H)$ and $u_0\in E, u_0\neq 0$. From (3.2) and assertion 3 of n^0. 2.1 we have

$$\mathbf{M}\{f\}u_0 = \mathbf{M}\{fu_0\} = \mathbf{M}\{\phi_E^*(f\cdot u_0)\} = \mathbf{M}\{(\phi_{\mathbf{C}}^* f)\cdot u_0\} = \mathbf{M}\{\phi_{\mathbf{C}}^* f\}\cdot u_0.$$

Hence, (3.2) holds for scalar-valued functions too. Now let $f\in\ker\phi_{\mathbf{C}}^*$. It is easy to see that $\phi_{\mathbf{C}}^*$ is a homomorphism of involution algebras (the involution is complex conjugation). Thus $|f|^2=f\cdot\bar{f}\in\ker\phi_{\mathbf{C}}^*$. Then $\mathbf{M}\{|f|^2\}=\mathbf{M}\{\phi_{\mathbf{C}}^*|f|^2\}=0$, so $f=0$. Thus $\ker\phi_{\mathbf{C}}^*=\{0\}$.

d') $\Rightarrow$ d). As before it may be assumed that (3.2) holds for scalar-valued functions. Let E be a Banach space and $v\in E'$. We have

$$(v,\mathbf{M}\{f\}) = \mathbf{M}\{(v,f)\} = \mathbf{M}\{\phi_{\mathbf{C}}^*(v,f)\} = \mathbf{M}\{(v,\phi_E^* f)\} = (v,\mathbf{M}\{\phi_E^* f\}).$$

Since $v\in E'$ is arbitrary we obtain the required assertion.

b) $\Rightarrow$ d'). Due to the orthogonality of characters we have for $f\in\mathrm{Trig}(H)$,

$$f = \mathbf{M}\{f\} + \sum_{\substack{\chi\in H' \\ \chi\neq 1}} c_\chi\cdot\chi$$

and, moreover, this representation is unique. Similarly,

$$\phi_E^* f = \mathbf{M}\{\phi_E^* f\} + \sum_{\substack{\theta\in G' \\ \theta\neq 1}} d_\theta\cdot\theta.$$

The homomorphism ϕ^* does not change constants. Therefore,

$$\phi_E^* f = \mathbf{M}\{f\} + \sum c_\chi \cdot \phi' \chi.$$

Since ϕ' is injective, $\phi'\chi = 1$ at the right-hand side of this equality. Comparing these two representations for $\phi_E^* f$ we obtain (3.2) for $f \in \mathrm{Trig}(H)$. Then (3.2) is true for all $f \in \mathrm{CAP}(H)$ by Proposition 1.3, and d') holds with $E = \mathbf{C}$. □

The case $G = \mathbf{R}^n$, $H = \mathbf{T}^N = \mathbf{R}^N / (2\pi \mathbf{Z}^N)$ the N-dimensional torus. A homomorphism $\phi: \mathbf{R}^n \to \mathbf{T}^N$ can be included in a commutative diagram

$$\begin{array}{ccc} & & \mathbf{R}^n \\ & \overset{A}{\nearrow} & \downarrow p \\ \mathbf{R}^n & \xrightarrow[\phi]{} & \mathbf{T}^n \end{array}$$

where p is the canonical epimorphism and A is a linear map. The latter is defined by the formula

$$Ax = (\xi^1 \cdot x, \ldots, \xi^N \cdot x), \quad \xi^j \in \mathbf{R}^n, \; j = 1, \ldots, N,$$

where $\xi \cdot x = \xi_1 \cdot x_1 + \cdots + \xi_n \cdot x_n$. The dual homomorphism

$$\phi' : \mathbf{Z}^N = (\mathbf{T}^N)' \to \mathbf{R}^n = (\mathbf{R}^n)'$$

is defined by the formula

$$\phi' z = z_1 \xi^1 + \cdots + z_N \xi^N \in \mathbf{R}^n,$$

where $z = (z_1, \ldots, z_N) \in \mathbf{Z}^N$. Assertion b) of Theorem 3.1 means that in this case the vectors $\xi^1, \ldots, \xi^N$ are linearly independent over $\mathbf{Z}$, or, what is the same, over $\mathbf{Q}$. As a result we obtain the Kronecker-Weyl theorem in the usual form.

COROLLARY 3.2. *If a continuous function $f(y_1, \ldots, y_N)$ is 2π-periodic in each variable $y_j \in \mathbf{R}, j = 1, \ldots, N$, and if the vectors $\xi^1, \ldots, \xi^N$ are linearly independent over $\mathbf{Z}$, then the function $F(x) = f(\xi^1 \cdot x, \ldots, \xi^N \cdot x)$, $x \in \mathbf{R}^n$ is a.p. in the sense of Bohr, and*

$$\mathbf{M}\{F\} = (2\pi)^{-N} \int_0^{2\pi} \cdots \int_0^{2\pi} f(y) dy. \tag{3.3}$$

The operator ϕ^* is defined only on Bohr a.p. functions and, generally speaking, cannot be extended to Besicovitch a.p. functions. However, in one very important case such an extension is possible. Let the equivalent assertions of Theorem 3.1 be fulfilled. Then for $f \in \mathrm{CAP}(H; E)$ the following obvious equality holds:

$$\phi^*(\|f\|_E^p) = \|\phi^* f\|_E^p,\ 1 \leqslant p < \infty.$$

From Theorem 3.1 we obtain

$$\mathbf{M}\{\|\phi^* f\|_E^p\} = \mathbf{M}\{\phi^* \|f\|_E^p\} = \mathbf{M}\{\|f\|_E^p\},$$

that is, for $f \in \mathrm{CAP}(H; E)$,

$$\|\phi^* f\|_{B^p} = \|f\|_{B^p},\ 1 \leqslant p < \infty. \tag{3.4}$$

Thus, in our case the operator ϕ^* can be continuously extended to an isometric operator

$$\phi^*: B^p(H; E) \to B^p(G; E),\ 1 \leqslant p < \infty.$$

Under these conditions ϕ^* is also an isometric operator in the CAP-norm:

$$\|\phi^* f\|_{\mathrm{CAP}} = \|f\|_{\mathrm{CAP}},\ f \in \mathrm{CAP}(H; E). \tag{3.5}$$

Indeed, since $\phi(G)$ is dense in H, the ranges of the functions f and $\phi^* f = f \circ \phi$ have the same closure in E. This gives (3.5).

3.2. RELATIVE BOHR COMPACTIFICATIONS.

Consider an arbitrary subgroup

$$S \subset (G')_d = G'_B \tag{3.6}$$

and denote by $p_S: G_B \to S'$ the homomorphism dual to the embedding (3.6). Note that p_S is an epimorphism. Indeed, the injectivity of (3.6) gives $\overline{p_S(G_B)} = S'$, and the compactness of G_B implies that $p_S(G_B)$ is closed.

Henceforth we will denote the group S' by $G_{B,S}$ and refer to it as the *relative Bohr compactification* (relative to the subgroup S). Strictly speaking, the relative Bohr compactification is the pair $(G_{B,S}, i_{B,S})$, $i_{B,S} = p_{S'} \circ i_B$, or the pair $(G_{B,S}, p_S)$, but later on we will use this slightly imprecise terminology because it cannot lead to any misunderstanding. Composition of the canonical embedding $i_B: G \to G_B$ and the epimorphism p_S gives a homomorphism $i_{B,S}: G \to G_{B,S}$ with dense range. This homeomorphism is injective iff the subgroup S is dense in G'. Indeed, $i_{B,S}$ is dual to the composition $S \to (G')_d \to G'$, where the first arrow is the embedding (3.6) and the second is the identity map.

The considerations of n°. 3.1 can be applied to the epimorphism p_S. Consequently, the following operators are well-defined:

$$p_S^*: C(G_{B,S}; E) \to \mathrm{CAP}(G; E) = C(G_B; E), \tag{3.7}$$

$$p_S^*: L^p(G_{B,S}; E) \to B^p(G; E) = L^p(G_B; E), \quad 1 \leqslant p < \infty \tag{3.8}$$

(L^p is understood with respect to the Haar measure on the corresponding group). These operators are isometric (see (3.4), (3.5)) and therefore have closed ranges, which may be described in terms of spectra of a.p. functions. For any set $\mathfrak{S}$ of a.p. functions and for any $K \subset (G')_d$, put

$$\mathfrak{S}_K = \{f \in \mathfrak{S} \mid \mathrm{sp}(f) \subset K\}.$$

PROPOSITION 3.3. *The following hold:*

$$p_S^* \mathrm{Trig}(G_{B,S}; E) = \mathrm{Trig}_S(G: E), \tag{3.9}$$

$$p_S^* C(G_{B,S}; E) = \mathrm{CAP}_S(G; E), \tag{3.10}$$

$$p_S^* L^P(G_{B,S}; E) = B^p_S(G; E), \ 1 \leqslant p < \infty \tag{3.11}$$

Proof. Equality (3.9) is valid because $(G_{B,S})' = S'$ and the restriction of p_S^* to characters coincides with the embedding $S \subset G'$. Due to Theorem 2.4 and relation (2.16), $\mathrm{CAP}_S(G; E)$ and $B^p_S(G; E)$, $1 \leqslant p < \infty$, are the closure of $\mathrm{Trig}_S(G; E)$ in $\mathrm{CAP}(G; E)$ and $B^p(G; E)$, respectively. Since the range of p_S^* is closed, (3.9) together with the preceding implies (3.10) and (3.11). □

PROPOSITION 3.4. *The following assertions holu.*

$$\mathrm{CAP}(G; E) = \bigcup \mathrm{CAP}_S(G; E), \tag{3.12}$$

$$B^p(G; E) = \bigcup B^p_S(G: E), \ 1 \leqslant p < \infty, \tag{3.13}$$

where the union is taken over all countable subgroups $S \subset (G')_d$.

Proof. If $f \in \mathrm{CAP}(G; E)$ (respectively, $f \in B^p(G; E)$), then $f \in \mathrm{CAP}_{S(f)}(G; E)$ (respectively, $f \in B^p_{S(f)}(G; E)$), where $S(f)$ is the frequency group. It remains to apply Proposition 2.3. □

The following proposition is a version of Proposition 3.4 for functions with values in a metric space.

PROPOSITION 3.5. *Let Y be a metric space and $f: G \to Y$ an a.p. function. Then there is an at most countable subgroup $S \subset (G')_d$ such that f extends to a continuous map $\tilde{f}: G_{B,S} \to Y$, i.e. $f = \tilde{f} \circ i_{B,S}$.*

Proof. By Remark 1.4 we may assume without loss of generality that the space Y is compact (if necessary we can replace it by $\overline{f(G)}$). Consider the embedding $j: Y \to l^2$ [75]. Then $j \circ f$ is an a.p. function with values in l^2, and by Proposition 3.3, 3.4 it can be represented as $j \circ f = g \circ i_{B,S}$, where $g \in C(G_{B,S}; l^2)$, for some countable subgroup $S \subset (G')_d$. Since $j(Y)$ is closed, $g(G_{B,S}) \subset j(Y)$. Furthermore, the map $j^{-1}: j(Y) \to Y$ is well-defined and continuous on $j(Y)$. To get the desired extension we put $\tilde{f} = j^{-1} \circ g$. □

The minimal subgroup $S(f) \subset (G')_d$ with the property stated in Proposition 3.5 is called the *frequency group of the map f*.

REMARK 3.5. In the sequel it is especially important that for a countable group S the group $G_{B,S}$ is metrizable. This follows from the equality $(G_{B,S})' = S$ and the fact [87] that a compact abelian group metrizable iff its group of characters is countable.

The group $G_{B,S}$ is natural on S in the following sense. If $S_1 \subset S_2 \subset S_3$, then there is a commutative diagram

$$\begin{array}{ccc} & G_{B,S_3} & \\ \swarrow & & \searrow \\ G_{B,S_2} & \longrightarrow & G_{B,S_1} \end{array}$$

where the arrows are dual to the embeddings $S_i \subset S_j$, $i < j$. In particular, the groups $G_{B,S}$, where S runs over all countable subgroups of $(G')_d$, form a projective system [8]. It can be proved that

$$G_B = \lim \operatorname{proj} G_{B,S},$$

where S runs over all countable subgroups of $(G')_d$.

The case $G = \mathbf{R}^n$. Let $\xi^1, \ldots, \xi^N \in \mathbf{R}^n$ be rationally independent and let $S \subset \mathbf{R}^n$ be the subgroup generated by them. Thus $S \simeq \mathbf{Z}^N$, where the isomorphism is given by the formula $\mathbf{Z}^N \ni (z_1, \ldots, z_N) \mapsto \sum z_i \xi^i \in S$. Hence $\mathbf{R}^n_{B,S} = S' = \mathbf{T}^N$.

Now let Γ be the vector space over $\mathbf{Q}$ generated by the vectors $\xi^1, \ldots, \xi^N$. Then $\dim_{\mathbf{Q}} \Gamma = N$. The space Γ can be represented as the union of an increasing collection of subgroups isomorphic to $\mathbf{Z}^N$:

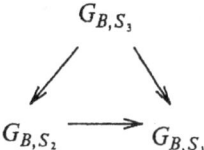

$$\sum_i \mathbf{Z}\cdot\xi^i \subset \sum_i (\mathbf{Z}/2)\cdot\xi^i \subset \cdots \subset \sum_i (\mathbf{Z}/k!)\cdot\xi^i \subset \cdots.$$

The dual maps give a projective system of N-dimensional tori:

$$\mathbf{T}^N \leftarrow \mathbf{T}^N \leftarrow \cdots \leftarrow \mathbf{T}^N \leftarrow \cdots.$$

Then

$$\mathbf{R}^n_{B,\Gamma} = \lim\operatorname{proj} \mathbf{T}^N.$$

The group $\mathbf{R}^n$ is the union of the family of subgroups S_γ isomorphic to $\mathbf{Z}^{N_\gamma}$. Therefore, $\mathbf{R}^n_B = \lim\operatorname{proj} T^{N_\gamma}$. Note that the dimensions N_γ of these tori are not bounded. Thus the space $\mathbf{R}^n_B$ has infinite dimension.

4. Stepanov almost periodic functions.

In this and the following sections we only consider functions defined on the real axis (for simplicity).

4.1. DEFINITION AND SIMPLE PROPERTIES.

First we introduce the space $BS^p(\mathbf{R};E)$, $1 \leqslant p < \infty$, of functions that are bounded (of exponent p) in the sense of Stepanov. It consists of the functions having finite norm

$$||f||^p_{S^p} = \sup_{t\in\mathbf{R}} \int_t^{t+1} ||f(\tau)||^p_E d\tau. \tag{4.1}$$

It is easy to check that $BS^p(\mathbf{R};E)$ is a Banach space with norm (4.1). An equivalent description may be given. For a function $f \in L^p_{loc}(\mathbf{R};E)$, put

$$f^b(t) = f(t+\eta), \quad \eta\in[0,1], \ t\in\mathbf{R}; \tag{4.2}$$

$f^b(t)$ is regarded as a function with values in the space $L^p(0,1;E)$*. Then

$$BS^p(\mathbf{R};E) = \{f\in L^p_{loc}(\mathbf{R};E) \mid f^b \in L^\infty(\mathbf{R};L^p(0,1;E))\}, \tag{4.3}$$

and, moreover,

* In this situation, the function f^b was introduced by Bochner [113], which is why we will call the map $f \mapsto f^b$ the Bochner transform.

$$||f||_{S^p} = ||f^b||_{L^\infty}. \tag{4.4}$$

Note that f^b is a continuous function (by construction) with values in $L^p(0, 1; E)$. Therefore, in (4.3) L^∞ may be replaced by the space C_b. It is easy to check that $BS^p(\mathbf{R}; E)$ is a Banach $C_b(\mathbf{R})$- module relative to the pointwise operations. Furthermore, there is a continuous embedding

$$BS^p(\mathbf{R}; E) \subset BS^q(\mathbf{R}; E), \quad q \leqslant p. \tag{4.5}$$

A function $f \in BS^p(\mathbf{R}; E)$, $1 \leqslant p < \infty$, is called an *almost periodic function in the sense of Stepanov* and of exponent p (abbreviated as S^p-a.p.) if $f^b \in CAP(\mathbf{R}; L^p(0, 1; E))$. In this case the ϵ-almost periods of the function f^b are called the ϵ-almost periods of f. The space of S^p-a.p. functions with values in E is denoted by $S^p(\mathbf{R}; E)$. Obviously, $CAP(\mathbf{R}; E) \subset S^p(\mathbf{R}; E)$. We given an equivalent description of the space of S^p-a.p. functions. For $f \in BS^p(\mathbf{R}; E)$ the function f^b may be interpreted as a function taking values in $BS^p(\mathbf{R}; E)$: all we need is to think that $\eta \in \mathbf{R}$ in definition (4.2). Then

$$S^p(\mathbf{R}; E) = \{f \in BS^p(\mathbf{R}; E) \,|\, f^b \in CAP(\mathbf{R}; BS^p(\mathbf{R}; E))\}.$$

Moreover, an exact description of the image of $S^p(\mathbf{R}; E)$ under the Bochner transform $f \mapsto f^b$ can be given. First note that $f \mapsto f^b$ is isometric in view of (4.4). Introduce the operators $T_\tau : \phi(\cdot) \mapsto \phi(\cdot + \tau)$ of left shift. They form a group of isometric operators in $BS^p(\mathbf{R}; E)$ (which is not strongly continuous). Denote by $BS^p(\mathbf{R}; E)^b \subset C_b(\mathbf{R}; BS^p(\mathbf{R}; E))$ the image of $BS^p(\mathbf{R}; E)$, and by $S^p(\mathbf{R}; E)^b \subset CAP(\mathbf{R}; BS^p(\mathbf{R}; E))$ the image of $S^p(\mathbf{R}; E)$, under the Bochner transform. The following equality holds:

$$BS^p(\mathbf{R}; E)^b = \{g \in C_b(\mathbf{R}; E)) \,|\, g(t+\tau) = (T_\tau g)(t)\}. \tag{4.6}$$

(Here the operators T_τ act on the range of the function g.) Indeed, for $g = f^b, f \in BS^p(\mathbf{R}; E)$, the equality $g(t+\tau) = (T_\tau g)(t)$ is obvious. Conversely, if g lies in the right-hand side of (4.6), then $g = f^b$ for $f = g(0) \in BS^p(\mathbf{R}; E)$.

PROPOSITION 4.1. *The space $S^p(\mathbf{R}; E)$ is closed in* $BS^p(\mathbf{R}; E)$

Proof. Since the transformation ^{b} is isometric, a subspace $BS^p(\mathbf{R}; E)^b$ is closed in $C_b(\mathbf{R}; BS^p(\mathbf{R}; E))$. Also, $S^p(\mathbf{R}; E)^b = BS^p(\mathbf{R}; E)^b \cap CAP(\mathbf{R}; BS^p(\mathbf{R}; E))$ and is therefore closed in $BS^p(\mathbf{R}; E)^b$. Since the Bochner transform is isometric we obtain the required assertion. □

The Bochner transform also shows that $S^p(\mathbf{R}; E)$ is a Banach $CAP(\mathbf{R})$-module and that for an arbitrary bounded linear operator $A : E \to F$ there is an induced homomorphism $A_* : S^p(\mathbf{R}; F) \to S^p(\mathbf{R}; E)$, $(A_* f)(t) = Af(t)$, of Banach modules. Then

(4.5) implies the existence of a continuous embedding

$$S^p(\mathbf{R}; E) \subset S^q(\mathbf{R}; E), \quad q \leqslant p. \tag{4.7}$$

(By the approximation theorem (Theorem 4.4) it is dense.)

The following theorem is an S^p-version of Bochner's criterion.

THEOREM 4.2. *For $f \in BS^p(\mathbf{R}; E)$ the following assertions are equivalent:*

a) $f \in S^p(\mathbf{R}; E)$;

b) the set of translates $\{T_\tau f\} = \{f(\cdot + \tau), \tau \in \mathbf{R}$, is relatively compact in $\mathrm{BS}^p(\mathbf{R}; E)$.

Proof. It suffices to apply Bochner's ordinary criterion to the function f^b and to note that $f^b(t) = (T_t f^b)(0) = T_t f$. □

4.2. APPROXIMATION THEOREM.

First we give a description of the image $\mathrm{Trig}(\mathbf{R}; E)^b$ of the space $\mathrm{Trig}(\mathbf{R}; E)$ under the Bochner transform.

PROPOSITION 4.3. *The following equality holds:*

$$\mathrm{Trig}(\mathbf{R}; E)^b = \mathrm{BS}^p(\mathbf{R}; E)^b \cap \mathrm{Trig}(\mathbf{R}; \mathrm{BS}^p(\mathbf{R}; E)). \tag{4.8}$$

Proof. Let

$$\phi(t) = \sum \phi_k e^{i\lambda_k t} \in \mathrm{Trig}(\mathbf{R}; E).$$

Then

$$\phi^b(t) = \sum (\phi_k e^{i\lambda_k \eta}) e^{i\lambda_k t}.$$

Hence it follows that

$$\mathrm{Trig}(\mathbf{R}; E)^b \subset \mathrm{BS}^p(\mathbf{R}; E)^b \cap \mathrm{Trig}(\mathbf{R}; \mathrm{BS}^p(\mathbf{R}; E)).$$

Conversely, let $g = \sum g_k e^{i\lambda_k t}$, where $g_k \in \mathrm{BS}^p(\mathbf{R}; E)$ is taken from the right-hand side of (4.8). Then

$$g(t+\tau) = \sum (g_k e^{i\lambda_k \tau}) e^{i\lambda_k t}.$$

But

$$(T_\tau g)(t) = \sum (T_\tau g_k) e^{i\lambda_k t}.$$

By (4.6) this gives

$$(T_\tau g_k)(x) = g_k(x+\tau) = g_k(x)e^{i\lambda_k\tau}.$$

Consequently, $g_k(x) = g_k(0)\exp(i\lambda_k x)$. Then $g = \phi^b$ for

$$\phi(t) = \sum \phi_k \exp(i\lambda_k t), \ \phi_k = g_k(0). \ \square$$

Now we establish the approximation theorem:

THEOREM 4.4. *The space* Trig($\mathbf{R}$; E) *is dense in* $S^p(\mathbf{R}; E)$.

Proof. Let $f \in S^p(\mathbf{R}; E)$. Consider the function f^b and apply the Bochner-Fejér operators (see n°. 2.3):

$$(L_\gamma f^b)(x) = \mathbf{M}\{\lambda_\gamma(y) f^b(x-y)\}.$$

Since $S^p(\mathbf{R}; E)^b$ is a closed and translation-invariant subspace of CAP($\mathbf{R}$; $BS^p(\mathbf{R}; E)$), we have $L_\gamma f^b \in S^p(\mathbf{R}; E)$. Hence $L_\gamma f^b \in BS^p(\mathbf{R}; E)^b \cap \mathrm{Trig}(\mathbf{R}; BS^p(\mathbf{R}; E)) = \mathrm{Trig}(\mathbf{R}; E)^b$, that is, $L_\gamma f^b = \phi_\gamma^b$, where $\phi_\gamma \in \mathrm{Trig}(\mathbf{R}; E)$. Then note that $f^b = \lim L_\gamma f^b$ in CAP($\mathbf{R}$; $BS^p(\mathbf{R}; E)$) and that the Bochner transform is an isometry. Therefore $f = \lim \phi_\gamma$ in $S^p(\mathbf{R}; E)$. $\square$

PROPOSITION 4.5. *The natural embedding* $\mathrm{CAP}(\mathbf{R}; E) \subset B^p(\mathbf{R}; E)$, $1 \leqslant p < \infty$, *can be continuously extended to an embedding* $S^p(\mathbf{R}; E) \subset B^p(\mathbf{R}; E)$.

Proof. There is an obvious estimate for $f \in \mathrm{CAP}(\mathbf{R}; E)$,

$$\lim_{T\to\infty} \frac{1}{2T} \int_{-T}^{T} ||f(\tau)||_E^p d\tau \leqslant \sup_{t\in\mathbf{R}} \int_t^{t+1} ||f(\tau)||_E^p.$$

In other words,

$$||f||_{B^p} \leqslant ||f||_{S^p}, \ f \in \mathrm{CAP}(\mathbf{R}; E).$$

This proves the statement, because the spaces CAP($\mathbf{R}$; E) and Trig($\mathbf{R}$; E) are dense in $S^p(\mathbf{R}; E)$. $\square$

4.3. MEAN VALUE.

By Proposition 4.5 the mean value $\mathbf{M}$ is defined for S^p- a.p. functions. Moreover, formula (2.2) for the mean value remains true in this case too.

THEOREM 4.6. *For* $f \in S^1(\mathbf{R}; E)$ *we have*

$$\mathbf{M}\{f\} = \lim_{T\to\infty} \frac{1}{2T} \int_{a-T}^{a+T} f(\tau)d\tau, \tag{4.9}$$

where the limit is uniform in $a \in \mathbf{R}$.

Proof. Let $f_n \in \mathrm{CAP}(\mathbf{R}; E)$ and $f = \lim f_n$ in $S^1(\mathbf{R}; E)$. Put

$$c(T, a, n) = \frac{1}{2T} \int_{a-T}^{a+T} f_n(\tau)d\tau.$$

It is easy to see that

$$\frac{1}{2T} \int_{a-T}^{a+T} f(\tau)d\tau = \lim_{n\to\infty} c(T, a, n)$$

uniformly in a and T. This justifies the permutation of limits in the equality

$$\mathbf{M}\{f\} = \lim_{n\to\infty} \lim_{T\to\infty} c(T, a, n)$$

and leads to formula (4.9). □

Now we can define the Bochner-Fejér operators either by restriction from the space of Besicovitch a.p. functions or by formula (2.9):

$$(L_\gamma f)(t) = \mathbf{M}_\tau\{\lambda_\gamma(\tau)f(t-\tau)\} = \mathbf{M}_\tau\{\lambda_\gamma(t-\tau)f(\tau)\}.$$

A straightforward calculation shows that the operators L_γ commute with the Bochner transform:

$$L_\gamma f^b = (L_\gamma f)^b.$$

This fact, together with the Theorem 2.4, implies:

PROPOSITION 4.7. *Let* $\mathfrak{F} \subset S^p(\mathbf{R}; E)$, $1 \leqslant p < \infty$, *be a relatively compact set. Then* $\lim L_\gamma f = f$ *uniformly for* $f \in \mathfrak{F}$ *in the topology of this space.* □

Due to the inclusion $S^p(\mathbf{R}; E) \subset B^p(\mathbf{R}; E)$ the Fourier-Bohr transform and the spectrum are defined for S^p-a.p.functions too. The spectrum of f is at most countable and $\mathrm{sp}(f) = \mathrm{sp}(f^b)$.

5. Weakly almost periodic functions.

Like any other "theory of vector functions" the theory of a.p. functions admits a weak version. In this section the main properties of weakly a.p. functions are presented. As in Section 4, in order to simplify arguments functions on $\mathbf{R}$ will be considered only.

5.1. MAIN PROPERTIES.

A function $f\colon \mathbf{R}\to E$, where E is a Banach space, is called weakly a.p. if for any $v\in E'$ the scalar function $(v, f(t))$ is almost periodic in the sense of Bohr. All such functions constitute a vector space, and even a module, over CAP($\mathbf{R}$), denoted by $\mathrm{CAP}_w(\mathbf{R}; E)$, which can be endowed with the topology generated by the seminorms

$$p_v(f) = \sup_{t\in\mathbf{R}} |(v, f(t))|, \quad v\in E'$$

(the topology of weak uniform convergence).

Here we give some simplest properties of weakly a.p. functions.

1. Any a.p. function is weakly a.p.

2. If $f\in\mathrm{CAP}_w(\mathbf{R}; E)$, then it is bounded and the set $f(\mathbf{R})$ is separable. In particular, $f\in L^\infty(\mathbf{R}; E)$.

(*Proof.* For any $u\in E'$ the function $(u, f(t))$ is a.p., and consequently bounded. By the Banach-Steinhaus theorem this implies the boundedness of f.

To prove the second assertion we first note that f is weakly continuous. Thus, for any $t_0\in\mathbf{R}$ there is a sequence $\{r_k\}\in\mathbf{Q}$ such that $\lim r_k = t_0$ and $w\text{-}\lim f(r_k) = f(t_0)$. By Mazur's theorem [15] there is a sequence of convex combinations of the elements $f(r_k)$ with rational coefficients which is strongly convergent to $f(t_0)$. This gives the required statement. □)

3. If the net $\{f_\gamma\}\subset\mathrm{CAP}_w(\mathbf{R}; E)$ is weakly uniformly convergent on $\mathbf{R}$ to f, then $f\in\mathrm{CAP}_w(\mathbf{R}; E)$.

*4. Let $f\in\mathrm{CAP}_w(\mathbf{R}; E)$ and suppose the net $\{f(t+s_\gamma)\}$ is weakly uniformly convergent to $g(t)$. Then**

$$\overline{\mathrm{co}}\, f(\mathbf{R}) = \overline{\mathrm{co}}\, g(\mathbf{R}), \tag{5.1}$$

$$\sup_{t\in\mathbf{R}} \| f(t) \| = \sup_{t\in\mathbf{R}} \| g(t) \|. \tag{5.2}$$

(*Proof.* Let $v\in\mathrm{co}\, g(\mathbf{R})$, i.e.

$$v = \sum_{k=1}^{N} \alpha_k g(t_k), \quad t_1, \ldots, t_N\in\mathbf{R},\ \alpha_k \geqslant 0,\ \sum\alpha_k = 1.$$

* By $\mathrm{co}\,A$ and $\overline{\mathrm{co}}\,A$ are denoted the convex hull of the set A and the closure of the convex hull of A, respectively.

Then

$$v = w\text{-lim}\sum \alpha_k f(t_k + s_\gamma) = w\text{-lim} v_\gamma,$$

where $v_\gamma \in \mathrm{co} f(\mathbf{R})$. By Mazur's theorem, $\overline{\mathrm{co}}\, g(\mathbf{R})$ is weakly closed. Hence $v \in \mathrm{co} f(\mathbf{R})$ and $\mathrm{co}\, g(\mathbf{R}) \subset \overline{\mathrm{co}} f(\mathbf{R})$.

Now observe that

$$w\text{-lim}\, g(t - s_\gamma) = f(t) \tag{5.3}$$

uniformly in $t \in \mathbf{R}$. Indeed, fix $u \in E'$. Then for any $\epsilon > 0$,

$$\sup_{t \in \mathbf{R}} |(u, f(t + s_\gamma)) - (u, g(t))| \leqslant \epsilon$$

for sufficiently large γ. However, for the same γ,

$$\sup_{t \in \mathbf{R}} |(u, f(t)) - (u, g(t - s_\gamma))| \leqslant \epsilon.$$

This proves (5.3).

Now in order to obtain the inclusion $\overline{\mathrm{co}} f(\mathbf{R}) \subset \overline{\mathrm{co}}\, g(\mathbf{R})$ it is sufficient to interchange f and g in the first part of the proof.

Let us prove (5.2). We have

$$||g(t)|| \leqslant \liminf ||f(t + s_\gamma)|| \leqslant \sup_{t \in \mathbf{R}} ||f(t)||,$$

because the norm in a Banach space is lower semicontinuous in the weak topology [15]. Interchanging f and g and using (5.3) we obtain the opposite inequality. □)

5.2. MEAN VALUE.

Let $f: \mathbf{R} \to E$ be a weakly a.p. function. The continuous linear functional $u \mapsto \mathbf{M}\{(u, f)\}$ is defined on the space E'; it is denoted by $\mathbf{M}\{f\} \in E''$. After that the Fourier-Bohr transform and the spectrum of an a.p. function can be defined. The Fourier-Bohr transform

$$\hat{f}(\xi) = \mathbf{M}\{f(t) e^{-i\xi \cdot t}\}$$

is, generally speaking, a function with values in E''. The spectrum of a weakly a.p. function is defined by the usual formula:

$$\mathrm{sp}(f) = \{\xi \in \mathbf{R} \mid \hat{f}(\xi) \neq 0\}.$$

It is easy to see that

$$\mathrm{sp}(f) = \bigcup_{u \in E'} \mathrm{sp}((u, f)).$$

The uniqueness theorem remains true for a.p. functions:

PROPOSITION 5.1. *Let $f: \mathbf{R} \to E$ be a weakly a.p. function and $\hat{f}(\xi) \equiv 0$. Then $f \equiv 0$.* □

Now recall that a sequence $\{V_n\}$ is called weakly fundamental if for any $u \in E'$ the sequence $\{(u, V_n)\}$ is convergent. A Banach space E is called *weakly sequentially complete* if any weakly fundamental sequence of its elements is weakly convergent to some element of E. Any reflexive space is weakly sequentially complete.

PROPOSITION 5.2. *If E is weakly sequentially complete and $f \in \mathrm{CAP}_w(\mathbf{R}; E)$, then $\mathbf{M}\{f\} \in E$ and*

$$\mathbf{M}\{f\} = w\text{-}\lim_{T \to \infty} \frac{1}{2T} \int_{a-T}^{a+T} f(t)dt, \tag{5.4}$$

where the limit is uniformly in $a \in \mathbf{R}$.

Proof. By statement 2 of n°. 5.1, $f \in L^\infty(\mathbf{R}; E)$ and hence f is locally integrable. Therefore the integrals in (5.4) make sense. Further, for $u \in E'$,

$$(u, \mathbf{M}\{f\}) = \lim \frac{1}{2T} \int_{a-T}^{a+T} (u, f(t))dt = \lim \left(u, \frac{1}{2T} \int_{a-T}^{a+T} (f(t)dt)\right).$$

By letting T tend to infinity along some sequence we obtain (5.4), since E is weakly sequentially complete. □

PROPOSITION 5.3. *Let E be a weakly sequentially complete space. Then the spectrum of any weakly a.p. function $f: R \to E$ is at most countable.**

Proof. By virtue of the separability of the image of f the space E may be regarded being separable too. Choose some countable set of functionals $\{u_n\}$ from E' such that $||u_n||_{E'} = 1$ and for any $v \in E$,

$$||v||_E = \sup_n |(u_n, v)|.$$

It is sufficient to take some countable and everywhere dense subset of the unit sphere $\{v \in E \mid ||v||_E = 1\}$ of E and to construct, using the Hahn-Banach

* M.I. Kadets and K. Kyursten [26] has proved that the assumption of weak sequential completeness is unnecessary.

theorem, a sequence $u_n \in E'$, $||u_n||_{E'}=1$, such that $(u_n, v_n)=1$. Then we have

$$||\hat{f}(\xi)||_E = \sup_n |(u_n, \hat{f}(\xi))|$$

(by Proposition 5.2, $\hat{f}(\xi)\in E$). This implies that

$$\mathrm{sp}(f) = \bigcup_n \mathrm{sp}((u_n, f)).$$

But at the right-hand side stands a countable union of at most countable sets. □

The Bochner-Fejér operators can be defined on $\mathrm{CAP}_w(\mathbf{R}; E)$ because formula (2.9) makes sense:

$$(L_\gamma f)(t) = \mathbf{M}_\tau\{\lambda_\gamma(\tau)f(t-\tau)\} = \mathbf{M}_\tau\{\lambda_\gamma(t-\tau)f(\tau)\}.$$

In general, $L_\gamma f\in \mathrm{Trig}(\mathbf{R}; E'')$. But if E is weakly sequential complete, then $L_\gamma f\in \mathrm{Trig}(\mathbf{R}; E)$ by Proposition 5.2. Moreover, the operators L_γ commute with continuous linear functionals: for a weakly a.p. f and $u\in E'$,

$$L_\gamma(u, f) = (u, l_\gamma f).$$

Together with Theorem 2.4 this gives

PROPOSITION 5.4. *If E is weakly sequentially complete and $f: \mathbf{R}\to E$ is a weakly a.p. function, then $f(t)=w-\lim(L_\gamma f)(t)$ uniformly on* $\mathbf{R}$.

5.3. EXTENTION TO THE BOHR COMPACTIFICATION.

We now consider the generalization of weakly a.p. functions to the Bohr compactification. At this point the situation is more complicated as in the case of strongly a.p. functions. For any E there is a continuous linear functional κv on E' acting according to the rule $\kappa v: u\mapsto (u, v)$. It is well-known that this correspondence defines an isometric embedding $\kappa: E\to E''$. Henceforth the space E will often be identified with its image $\kappa E\subset E''$.

Now let $f: \mathbf{R}\to E$ to be a weakly a.p. function. Then for any $u\in E'$ the function $(u, f(t))$ can be extended to a continuous function $\tilde{f}(t, u)$ on $\mathbf{R}_B$. It is easy to see that for any $t\in \mathbf{R}_B$ the map $u\mapsto \tilde{f}(t; u)$ is a continuous linear functional on E'. Therefore the function $\tilde{f}: \mathbf{R}_B\to E''$ is defined and $\tilde{f}|_{\mathbf{R}}=f$. Moreover, the function $\tilde{f}$ is obviously continuous in the E'-topology of the space E''. (This topology is generated by the system of seminorms $\{p_u\}$, $u\in E'$, where $p_u(v)=|v(u)|$, $v\in E''$). If E is weakly sequentially complete, then E' is sequentially closed in the E'-topology of the space E''. But when restricted to E this topology coincides with the ordinary weak topology.

THEOREM 5.5. *Let E be weakly sequentially complete and let $f: \mathbf{R} \to E$ be a weakly a.p. function. Then there exists a unique weakly continuous extension $\tilde{f}: \mathbf{R}_B \to E$ of the function f.*

Proof. Since the spectrum sp(f) is countable, (see Proposition 3.3 and 3.4) for any countable subgroup $S \subset \mathbf{R}_d$ such that sp(f) $\subset S$, the functions $\tilde{f}(t; u)$ are the inverse images under the canonical epimorphism $p_s: \mathbf{R}_B \to \mathbf{R}_{B,S}$ of the functions $\tilde{f}_S(t, u)$, which are defined for $t \in \mathbf{R}_{B,S}$. This defines the functions $\tilde{f}_S: \mathbf{R}_{B,S} \to E''$ (continuous in the E'-topology) such that $\tilde{f}_s \mid_{i_{B,S}(\mathbf{R})} = f$. Since $\mathbf{R}_{B,S}$ is metrizable and $i_{B,S}(\mathbf{R})$ is a dense subset of $\mathbf{R}_{B,S}$, for any $t \in \mathbf{R}_{B,S}$ there is a sequence $t_n \in \mathbf{R}_n$ such that $\lim i_{B,S}(t_n) = t$. Hence $\tilde{f}_S(t) = E'$-$\lim f(t_n) \in E$. Moreover, f_S is weakly continuous. This gives the required assertion. □

REMARK 5.6. It is important to note that $\tilde{f}(t) = \tilde{f}_S(p_S t)$, where $S \subset \mathbf{R}_d$ is a countable subgroup and $\tilde{f}_S: \mathbf{R}_{B,S} \to E$ is a weakly continuous extension of f to $\mathbf{R}_{B,S}$.

Theorem 5.5. implies the important

PROPOSITION 5.7. *If E is weakly sequentially complete, then there is a natural (algebraic) embedding* $\mathrm{CAP}_w(\mathbf{R}; E) \subset B^\infty(\mathbf{R}; E)$ *which coincides with the standard one on* $\mathrm{CAP}(\mathbf{R}; E)$.

Proof. By assertion 2 of n°. 5.1 we may assume that E is separable. Since $\tilde{f}$ is weakly continuous, $\tilde{f}$ is weakly measurable on $\mathbf{R}_B$. By virtue of the separability of E this function is then also strongly measurable. Identifying f with $\tilde{f}$ we obtain the required embedding. □

Bochner's criterion is valid for weakly a.p. functions too.

THEOREM 5.8. *Let E be a weakly sequentially complete space and let $f: \mathbf{R} \to E$ be a bounded weakly continuous function. Then the following two assertions are equivalent:*

a) $f \in \mathrm{CAP}_w(\mathbf{R}; E)$;

b) for any net $\{\tau_\gamma\} \subset \mathbf{R}$ there is a subnet $\{\tau_{\gamma'}\}$ such that the functions $f(\cdot + \tau_{\gamma'})$ are weakly convergent, uniformly on $\mathbf{R}$.

Proof. The implication b) ⇒ a) follows from the ordinary Bochner criterion.

a) ⇒ b). Let $f \in \mathrm{CAP}_w(\mathbf{R}; E)$. By virtue of the compactness of $\mathbf{R}_B$ there is a subnet $\{\tau_{\gamma'}\}$ such that $\lim \tau_{\gamma'} = \tau_0 \in \mathbf{R}_B$ (in the $\mathbf{R}_B$-topology). Then w-$\lim f(t + t_{\gamma'}) = w$-$\lim \tilde{f}(t + \tau_\gamma) = \tilde{f}(t + \tau_0)$ uniformly in $t \in \mathbf{R}$. □

The next assertion is often very useful [100].

PROPOSITION 5.9. *Let $E_0 \subset E_1$ be a continuous dense embedding of Banach spaces and let $f \in L^\infty(\mathbf{R}; E_0)$ be a weakly continuous function. If $f \in \mathrm{CAP}_w(\mathbf{R}; E_1)$, then $f \in \mathrm{CAP}_w(\mathbf{R}; E_0)$.*

Proof. Consider the dual embedding $E_1' \subset E_0'$. It is continuous and dense, therefore for any $v \in E_0'$ there is a sequence $v_n \in E_1'$ such that $v = \lim v_n$ in E_0'. We have the obvious estimate

$$|(v,f(t))-(v_n,f(t))| \leqslant ||f||_{L^\infty(\mathbf{R};E_0)}\cdot||v-v_n||_{E_0'}.$$

Since all (v_n, f) are a.p. this gives the required assertion. □

5.4. CRITERIA FOR STRONG ALMOST PERIODICITY.

Often it is very important to establish that some weakly a.p. function is in fact strongly a.p. The main such criterion is the following statement.

THEOREM 5.10. *A weakly a.p. function $f: \mathbf{R} \to E$ is almost periodic iff the set $f(\mathbf{R})$ is relatively compact in E.*

Proof. Necessity is obvious. We prove sufficiency.

Consider the E'-continuous extension $\tilde{f}: \mathbf{R}_B \to E''$ of the function f. First we show that $\tilde{f}(\mathbf{R}) \subset E$ (recall that E is considered to be embedded in E'' with the aid of the canonical isometry κ) and hence $\tilde{f}$ is weakly continuous on $\mathbf{R}_B$. Let $t_0 \in \mathbf{R}_B$ and let $\{t_\gamma\}$ be a net in $\mathbf{R}$ such that $\lim t_\gamma = t_0$. By the relative compactness of $f(\mathbf{R})$ we may assume (taking a subnet if necessary) that $\lim f(t_\gamma) = v \in E$ exist. From the E'-continuity of $\tilde{f}$ and the fact that $\tilde{f}_{\mathbf{R}} = f$ we obtain that $v = \lim f(t_\gamma) = E' - \lim f(t_\gamma) = \tilde{f}(t_0) \in E$. Thus $\tilde{f}(\mathbf{R}_B) \subset E$.

The same reasoning shows that $\mathrm{cl} f(\mathbf{R}) = w - \mathrm{cl} f(\mathbf{R})$ (cl and $w-$cl denote strong and weak closure, respectively). Moreover, $w - \mathrm{cl} f(\mathbf{R}) \supset \tilde{f}(\mathbf{R}_B)$. Hence $\tilde{f}(\mathbf{R}_B)$ is relatively compact.

We now establish the continuity of $\tilde{f}$ in the strong topology. If $\{t_\gamma\}$ is a convergent net in $\mathbf{R}_B$, $\lim t_\gamma = t_0$, then $\{\tilde{f}(t_\gamma)\}$ has at least one strong limit point (by the relative compactness of the image). But in view of the weak continuity of $\tilde{f}$, the net $\{\tilde{f}(t_\gamma)\}$ is weakly convergent to $\tilde{f}(t_0)$. Hence $\tilde{f}(t_0)$ is the unique strong limit point of $\{\tilde{f}(t_\gamma)\}$ and the continuity of f is proved. □

Let us give another criterion for strong almost periodicity. Put $f_x(t) = \tilde{f}(x+t)$, $x \in \mathbf{R}_B$, $t \in \mathbf{R}$.

PROPOSITION 5.11. *Let E be a weakly sequentially complete space and suppose the norm of E possesses the property:*

(H) if w-$\lim v_n = V$ and $\lim ||v_n|| = ||v||$, then $\lim v_n = v$.

Assume that a function $f: \mathbf{R} \to E$ is weakly a.p. and that for any $x \in \mathbf{R}_B$ the functions $||f_x(t)||$ are almost periodic. Then f is an a.p. function.

Proof. We establish the continuity of $\tilde{f}$ on $\mathbf{R}_B$. By remark 5.6 we consider the function $\tilde{f}_S$ on $\mathbf{R}_{B,S}$ instead of $\tilde{f}$ (here S is the frequency group of f). Incidentally, the norms of all functions $f_{S,y}(t) = \tilde{f}_S(y+t)$, $y \in \mathbf{R}_{B,S}$, $t \in \mathbf{R}$, are almost periodic, because $\tilde{f}(x) = \tilde{f}_S(p_s x)$ and the embeddings of $\mathbf{R}$ into Bohr compactifications commute with the canonical epimorphisms.

In proving the continuity of $\tilde{f}_S$ we need only consider sequences instead of nets (because $\mathbf{R}_{B,S}$ is metrizable). Moreover, by property (H) and the weak continuity of

$\tilde{f}_S$

it suffices to prove that $\tilde{f}_S$ is continuous in order to establish the continuity of the functions $\| f_S(y) \|_E$ on $\mathbf{R}_{B,S}$.

Consider now the functions $\| f_{S,y}(t) \|_E$. By their almost periodicity they may be extended to continuous functions ϕ_y on $\mathbf{R}_{B,S}, y \in \mathbf{R}_{B,S}$. The desired assertion will be proved if we establish the compatibility of the functions ϕ_y in the following sense:

$$\phi_y(t) = \phi(y+t), t \in \mathbf{R}, \tag{5.5}$$

where $\phi(x) = \phi_0(x)$ is the extension of $\| f(t) \|$ to $\mathbf{R}_{B,S}$.

So, let $y_n \in \mathbf{R}$, $\lim y_n = y \in \mathbf{R}_{B,S}$. Then for $t \in \mathbf{R}$, $\phi(y+t) = \lim \phi(y_n+t) \geqslant \| w - \lim f(y_n+t) \| = \| \tilde{f}_S(y+t) \| = \phi_y(t)$ i.e.

$$\phi(y+t) \geqslant \phi_y(t),\ t \in \mathbf{R},\ y \in \mathbf{R}_{B,S}. \tag{5.6}$$

Similarly,

$$\phi_y(t-y) = \lim \phi_y(t-y_n) \geqslant \phi(t),\ t \in \mathbf{R}.$$

It readily follows that

$$\phi_y(y_n+t-y) \geqslant \phi(y_n+t).$$

Taking the limit we obtain

$$\phi_y(t) \geqslant \phi(y+t), \tag{5.7}$$

because the functions ϕ_y and ϕ are continuous on $\mathbf{R}_{B,S}$. Inequalities (5.6) and (5.7) gives (5.5) and the assertion is proved. □

The conditions of the last Proposition are fulfilled automatically for uniformly convex spaces. Recall that a Banach space E is uniformly convex if the conditions $\| u_n \| = \| v_n \| = 1$, $\| u_n + v_n \| \to 2$ imply that $\| u_n - v_n \| \to 0$. Any uniformly convex space is reflexive and, thus, weakly sequentially complete. Property (H) is also valid [18]. Moreover, it is known [18] that in any separable Banach space we may introduce an equivalent norm possessing property (H).

5.5. WEAK ALMOST PERIODICITY IN THE SENSE OF STEPANOV.

We will say that a function $f \in BS^p(\mathbf{R}; E)$ is weakly a.p. of exponent p in the sense of Stepanov (is weakly $S^p(\mathbf{R})$) if for any $v \in E'$ the function (V, f) lies in $S^p(\mathbf{R})$. The weakly S^p-a.p. functions with values in a Banach space E constitute the vector space $S^p_w(\mathbf{R}; E)$ (which is even a CAP(**R**)-module). Obviously, $\mathrm{CAP}_w(\mathbf{R}; E) \subset S^p_w(\mathbf{R}; E)$.

PROPOSITION 5.2. *Assume that the embedding* $E_0 \subset E_1$ *is dense and continuous. Then* $BS^p(\mathbf{R}; E_0) \cap S^p_w(\mathbf{R}; E_1) = S^p_w(\mathbf{R}; E_0)$.

The *proof* is similar to that of Proposition 5.9.

Consider the Bochner transform on weakly S^p-a.p. functions. Since by definition such functions belong to $BS^p(\mathbf{R}; E)$, their Bochner transforms in $L^\infty(\mathbf{R}; L^p(0, 1; E))$.

PROPOSITION 5.13. *Let E be a reflexive Banach space and* $1 < p < \infty$. *Then the inclusion*

$$S^p_w(\mathbf{R}; E) \subset \{f \in BS^p(\mathbf{R}; E) \mid f^b \in \mathrm{CAP}_w(\mathbf{R}; L^p(0, 1; E))\}$$

holds.

Proof. Since E is reflexive, $L^p(0, 1; E)' = L^{p'}(0, 1; E')$ for $p > 1$ [95]. Hence we need only show that for any $g \in L^{p'}(0, 1; E')$ the function

$$f_g(t) = \int_0^1 (g(\tau), f^b(t, \tau))d\tau = \int_0^1 (g(\tau), f(t+\tau))d\tau$$

is almost periodic. Since $f^b \in L^\infty(\mathbf{R}; L^p(0, 1; E))$, without loss of generality we may assume that $g = \chi \cdot v$, where $v \in E'$ and χ is the characteristic function of some interval $[a, b] \subset [0, 1]$. Then

$$f_g(t) = \int_a^b (v, f^b(t, \tau))d\tau. \tag{5.8}$$

But $(v, f) \in S^p(\mathbf{R})$ and, consequently, $(v, f^b) = (v, f)^b \in \mathrm{CAP}(\mathbf{R}; L^p(0, 1))$. This, together with (5.8), gives the required assertion. □

THEOREM 5.14. *Let E be a reflexive Banach space and* $1 < p < \infty$. *Then there exist a unique (algebraic) embedding* $S^p_w(\mathbf{R}; E) \subset B^p(\mathbf{R}; E)$ *which coincides with the standard one on* $S^p(\mathbf{R}; E)$. *Moreover,*

$$\| f \|_{B^p} \leqslant \| f \|_{S^p}, \; f \in S^p_w(\mathbf{R}; E). \tag{5.9}$$

Proof. Consider the Bochner transform $f^b(t, s) = f(t+s)$, $t, s \in \mathbf{R}$, for $f \in S^p_w(\mathbf{R}; E)$. By Proposition 5.13, f^b as a function of the first argument lies in $\mathrm{CAP}_w(\mathbf{R}; X)$, where $X = L^p(T_1, T_2; E)$ and the interval $[T_1, T_2]$ is arbitrary. Since X is reflexive, and thus weakly sequentially complete, by Theorem 5.5 f^b may be extended to an

X-valued function which is weakly continuous on $\mathbf{R}_B$ (we preserve the notation f^b for the extended function). In addition, by Proposition 5.7 we have $f^b \in B^\infty(\mathbf{R}; X) = L^\infty(\mathbf{R}_B; X)$. The function $f^b(t, s)$ is a measurable on $\mathbf{R}_B \times \mathbf{R}$ as a function of two variables, and, moreover,

$$f^b(t+\tau, s-\tau) = f^b(t, s), \quad (t, s) \in \mathbf{R}_B \times \mathbf{R}, \ \tau \in \mathbf{R} \tag{5.10}$$

(for $(t, s) \in \mathbf{R} \times \mathbf{R}$ this equality is obvious; the weak continuity of f^b on $\mathbf{R}_B$ implies the general case). In addition, this function is integrable on $\mathbf{R}_B \times [T_1, T_2]$ for any finite interval $[T_1, T_2]$. According to Fubini's theorem there is a subset $N \subset R$ of measure zero such that $f^b(t, s)$ is integrable with respect to $t \in \mathbf{R}_B$ for all $s \in \mathbf{R} \setminus N$. Choose some $s \in \mathbf{R} \setminus N$ and put

$$g(t) = f^b(t-s, s), \quad t \in \mathbf{R}_B. \tag{5.11}$$

By (5.10), g does not depend on $s \in \mathbf{R} \setminus N$. Moreover, the same equality (5.10) shows that (5.11) holds for all $s \in \mathbf{R}$, that is, we may assume $N = \varnothing$. The function g will be the desired extension of f^b to $\mathbf{R}_B$.

Now we will show that $g \in B^p(\mathbf{R}; E)$. Taking into account (5.11) and the invariance of the measure μ, by Tonelli's theorem we obtain

$$\int_{\mathbf{R}_B} d\mu(t) \int_0^1 \| f^b(t+s) \|_E^p ds = \int_{\mathbf{R}_B} d\mu(t) \int_0^1 \| g(t+s) \|_E^p ds =$$

$$= \int_0^1 ds \int_{\mathbf{R}_B} \| g(t+s) \|_E^p d\mu(t) = \int_0^1 ds \int_{\mathbf{R}_B} \| g(t) \|_E^p d\mu(t) = \mathbf{M}\{ \| g \|_E^p \}.$$

Since the left-hand side of the last equality is finite, this implies the embedding we need. The estimate (5.9) follows from the coincidence of the norm in B^p with the norm $<\cdot>_p$ defined be equality (2.8).

Due to the estimate (5.9) and Proposition 4.5 it suffices to check the coincidence of the embedding we have constructed with the standard one only on CAP($\mathbf{R}; E$). But this is obvious. □

6. Spaces of smooth almost periodic functions on $\mathbf{R}^n$.

The action of differential operators on functions defined on $\mathbf{R}^n$ gives rise to various spaces of "smooth" a.p. functions. Two cases are considered: the space of a.p. functions with classical smoothness and spaces of a.p. functions of Sobolev type. We restrict ourselves to the case of scalar functions. Incidently, all results automatically hold for functions with values in finite-dimensional spaces.

6.1. UNIFORM SMOOTHNESS.

Here and henceforth the standard notations will be used: $\partial^\alpha = \partial_1^{\alpha_1} \cdots \partial_n^{\alpha_n}$, where $\partial_j = \partial / \partial x_j$, $\alpha = (\alpha_1, \ldots, \alpha_n) \in \mathbf{Z}_+^n$ is a multi-index, $|\alpha| = \alpha_1 + \cdots + \alpha_n$. Put

$$C_b^m(\mathbf{R}^n) = \{f \in C_b(\mathbf{R}^n) \,|\, \partial^\alpha f \in C_b(\mathbf{R}^n),\ |\alpha| \leqslant m\},\ m \in \mathbf{Z}_+,$$

$$C_b^m(\mathbf{R}^n) = \bigcap_{m \in \mathbf{Z}_+} C_b^m(\mathbf{R}^n).$$

The space $C_b^m(\mathbf{R}^n)$ endowed with the norm

$$\|f\|_{(m)} = \sum_{|\alpha| \leqslant m} \sup_{x \in \mathbf{R}^n} |\partial^\alpha f(x)| \tag{6.1}$$

is a Banach space; $C_b^\infty(\mathbf{R}^n)$ with the seminorms $\|\cdot\|_{(m)}$, $m \in \mathbf{Z}_+$, is a Fréchet space.

We denote by $\mathrm{CAP}^m(\mathbf{R}^n)$, $m \in \mathbf{Z}_+$, the set of functions $f \in \mathrm{CAP}(\mathbf{R}^n)$ for which $\partial^\alpha f \in \mathrm{CAP}(\mathbf{R}^n)$ for $|\alpha| \leqslant m$, and put

$$\mathrm{CAP}^\infty(\mathbf{R}^n) = \bigcap_{m \in \mathbf{Z}_+} \mathrm{CAP}^m(\mathbf{R}^n).$$

It is easy to see that $\mathrm{CAP}^m(\mathbf{R}^n)$ is a closed subspace of $C_b^m(\mathbf{R}^n)$ for all $0 \leqslant m \leqslant \infty$.

Note that $\partial_j f \in \mathrm{CAP}(\mathbf{R}^n)$ if $f \in \mathrm{CAP}(\mathbf{R}^n)$ and if the derivative $\partial_j f$ is uniformly continuous on $\mathbf{R}^n$. Indeed,

$$h^{-1}[f(x + he_j) - f(x)] = h^{-1} \int_0^h \partial_j f(x + te_j)\,dt,$$

where e_j is a vector of the standard basis in $\mathbf{R}^n$. The left-hand side of this equality is clearly a.p. The uniform continuity guarantees the uniform convergence of the right- hand side to $\partial_j f(x)$ as $h \to 0$, and, consequently, gives $\partial_j f \in \mathrm{CAP}(\mathbf{R}^n)$. In particular we have

$$\mathrm{CAP}^\infty(\mathbf{R}^n) = \mathrm{CAP}(\mathbf{R}^n) \cap C_b^\infty(\mathbf{R}^n).$$

Now we consider the Bochner-Fejér operators. The following statement is valid:

PROPOSITION 6.1. *Let* $\mathfrak{F}$ *be an arbitrary relatively compact subset of* $\mathrm{CAP}^m(\mathbf{R}^n)$, $m \in \mathbf{Z}_+$. *Then* $\lim L_\gamma f = f$ *in* $\mathrm{CAP}^m(\mathbf{R}^n)$ *uniformly in* $f \in \mathfrak{F}$. *In particular, for* $f \in \mathrm{CAP}(\mathbf{R}^n)$ *we have* $\lim L_\gamma f = f$ *in the* $\mathrm{CAP}^\infty(\mathbf{R}^n)$*-topology.*

Proof. The commutativity of the operators L_γ with derivatives,

$$\partial^\alpha L_\gamma f = L_\gamma \partial^\alpha f,\ f \in \mathrm{CAP}^m(\mathbf{R}^n),\ |\alpha| \leqslant m,$$

proves the statement. □

This also gives that Trig($\mathbf{R}^n$) is dense in CAP^m($\mathbf{R}^n$) for all $0 \leqslant m \leqslant \infty$.

Differential operators and operators of multiplication by CAP$^\infty$-functions act on CAP$^\infty$($\mathbf{R}^n$) in a continuous manner. Thus, differential operators with CAP$^\infty$-coefficients, of the form

$$A = \sum_{|\alpha| \leqslant m} a_\alpha(x) D^\alpha, \tag{6.2}$$

are continuous on CAP$^\infty$($\mathbf{R}^n$). Here, as usual, $D^\alpha = D_1^{\alpha_1} \cdots D_n^{\alpha_n}$, where $D_j = i^{-1}\partial_j$. Recall that the formal adjoint of the operator A is defined by the formula

$$A^+ u = \sum_{|\alpha| \leqslant m} D^\alpha(\overline{a_\alpha(x)} u(x)). \tag{6.3}$$

Thus,

$$(Au, v) = (u, A^+ v) \tag{6.4}$$

for $u, v \in C_0^\infty(\mathbf{R}^n)$, where

$$(u, v) = \int_{\mathbf{R}^n} u(x)\overline{v(x)}dx$$

is the usual L^2-duality.

In the sequel an analogue of formula (6.4) for the Besicovitch duality

$$(u, v)_B = \mathbf{M}\{u \cdot \bar{v}\} \tag{6.5}$$

is very important.

PROPOSITION 6.2. *For $u, v \in \mathrm{CAP}^\infty(\mathbf{R}^n)$ the following formula holds:*

$$(Au, v)_B = (u, A^+ v)_B. \tag{6.6}$$

Proof. Consider the cube

$$K_T = \{x \in \mathbf{R}^n \mid |x_j| \leqslant T, j = 1, \ldots, n\}.$$

Integration by parts gives

$$\frac{1}{(2T)^n}\int_{K_T} Au\cdot\bar{v}\,dx - \frac{1}{(2T)^n}\int_{K_T} u\cdot\overline{A^+v}\,dx =$$

$$= \frac{1}{(2T)^n}\int_{\partial K_T}\sum_{|\alpha|+|\beta|\leqslant m} c_{\alpha\beta}(x,T)D^\alpha u\overline{D^\beta v}\,dx, \tag{6.7}$$

where $|c_{\alpha\beta}(x,T)|\leqslant C$, ($C$ does not depend on x and T). Since the $(n-1)$-dimensional measure of ∂K_T is equal $O(T^{n-1})$, the right-hand side of formula (6.7) tends to zero as $T\to\infty$ and we obtain (6.6). □

Proposition 6.2 allows us to extend differential operators by Besicovitch duality to various spaces of "generalized" a.p. functions in just the same way as formula (6.4) defines the action of differential operators on distributions. Also, it is useful to note that formula (6.6) is valid even for operators with CAP^m- coefficients.

6.2. SOBOLEV-BESICOVITCH SPACES.

For $k\in\mathbf{Z}_+$ and $1\leqslant p<\infty$ the space $W^{k,p}(\mathbf{R}^n_B)$ is defined as the completion of $CAP^\infty(\mathbf{R}^n)$ with respect to the norm

$$||f||^p_{k,p} = \sum_{|\alpha|\leqslant k} ||\partial^\alpha f||^p_{B^p}. \tag{6.8}$$

The notation $||\cdot||_{k,p}$ will be used also for the usual Sobolev norm - this will not lead to confusion. For integers $k<0$ and $1<p<\infty$, we define

$$W^{k,p}(\mathbf{R}^n_B) = [W^{-k,p'}(\mathbf{R}^n_B)]', \quad \frac{1}{p}+\frac{1}{p'} = 1. \tag{6.9}$$

Clearly, the space $W^{0,p}(\mathbf{R}^n_B)$ coincides with $B^p(\mathbf{R}^n)$. There are natural continuous and dense embeddings

$$W^{k,p}(\mathbf{R}^n_B)\subset W^{s,q}(\mathbf{R}^n_B),\ k\geqslant s,\ p\geqslant q. \tag{6.10}$$

For $k\geqslant s\geqslant 0$ this follows from the definition and form the properties of Besicovitch spaces. From this and (6.9) we obtain the desired properties in the case $0\geqslant k\geqslant s$. The case $s\leqslant 0\leqslant k$ reduces the previous two cases.

Also, it is useful to introduce the spaces

$$W^{\infty,p}(\mathbf{R}^n_B) = \bigcap_{k\in\mathbf{Z}} W^{k,p}(\mathbf{R}^n_B)$$

and

$$W^{-\infty,p}(\mathbf{R}^n_B) = \bigcup_{k\in\mathbf{Z}} W^{k,p}(\mathbf{R}^n_B).$$

These spaces are endowed with the natural topology. Observe that $W^{\infty,p}(\mathbf{R}^n_B) \not\subset \mathrm{CAP}^\infty(\mathbf{R}^n)$. The proof for the case $p=2$ is given in [89].

For $k\geqslant 0$ the spaces $W^{k,p}(\mathbf{R}^n_B)$ are obviously CAP^k-modules; moreover, multiplication is continuous in both variables simultaneously. For $\phi\in\mathrm{CAP}^k(\mathbf{R}^n)$ and $u\in W^{-k,p}(\mathbf{R}^n_B)$, $k>0$, multiplication is defined by the formula

$$(\phi u, v)_B = (u, \bar{\phi} v), \ \forall v\in W^{k,p'}(\mathbf{R}^n_B). \tag{6.11}$$

Here $(\cdot,\cdot)_B$ is the duality between $W^{-k,p}(\mathbf{R}^n_B)$ and $W^{k,p'}(\mathbf{R}^n_B)$; this agrees with (6.5). The operators of multiplication defined in this way commute with the natural embeddings.

The derivatives acting in $\mathrm{CAP}^\infty(\mathbf{R}^n)$ may be extended to continuous operators

$$D_j\colon W^{k,p}(\mathbf{R}^n_B) \to W^{k-1,p}(\mathbf{R}^n_B),\ j = 1,\ldots,n.$$

For $k\geqslant 1$ this is done by continuity, which follows from the definition of the norms (6.8), and for $k\leqslant 0$ this is done by duality:

$$(D_j u, v) = (u, D_j v)_B, \quad \forall v\in W^{1-k,p'}(\mathbf{R}^n_B). \tag{6.12}$$

The last definition is correct by Proposition 6.2. Thus, the action of a.p. differential operators is well-defined in Sobolev- Besicovitch spaces too. Any such m-order operator (for example with CAP^∞-coefficients) induces a continuous linear map

$$A\colon W^{k,p}(\mathbf{R}^n_B) \to W^{k-m,p}(\mathbf{R}^n_B).$$

Moreover,

$$(Au, v)_B = (u, A^+ v)_B,\ u\in W^{k,p}(\mathbf{R}^n_B),\ v\in W^{k-m,p'}(\mathbf{R}^n_B). \tag{6.13}$$

This follows from (6.6).

We can also define the Fourier-Bohr transform for functions from Sobolev-Besicovitch spaces:

$$\hat{f}(\xi) = (f(x), e^{i\xi\cdot x})_B,\ f\in W^{-\infty,p}(\mathbf{R}^n_B). \tag{6.14}$$

All formal properties of these transform hold in this case. In addition we have

$$\widehat{D_j f}(\xi) = \xi_j \hat{f}(\xi). \tag{6.15}$$

This formula is a version of the corresponding formula for the Fourier transform and follows from (6.12) and (6.14). The spectrum $\mathrm{sp}(f)$ and the frequency group for $f \in W^{-\infty,p}(\mathbf{R}_B^n)$ are defined in the usual manner. It is easy to see that $\mathrm{sp}(f)$ is at most countable in this case too. By formula (6.15), $\mathrm{sp}(D_j f) \subset \mathrm{sp}(f)$.

Now we consider the Bochner-Fejér operators. For $f \in W^{-\infty,p}(\mathbf{R}_B^n)$ we put

$$(L_\gamma f)(x) = (f(\cdot), \lambda_\gamma(x - \cdot))_B. \tag{6.16}$$

This formula makes sense because $\lambda_\gamma(x-\cdot) \in \mathrm{Trig}(\mathbf{R}^n)$ for any $x \in \mathbf{R}^n$. Using the eveness agreement on Bochner-Fejér kernels (Remark 2.5) it is easy to obtain the equality

$$(L_\gamma f, g)_B = (f, L_\gamma g)_B, \tag{6.17}$$

which, equivalently, may be taken as a definition of the operators L_γ on Sobolev-Besicovitch spaces.

PROPOSITION 6.3. *For $f \in W^{k,p}(\mathbf{R}_B^p)$, $-\infty \leqslant k \leqslant \infty$, $1 < p < \infty$, the equality $\lim L_\gamma f = f$ holds (in the topology of this space).*

Proof. It clearly suffices to consider the case of k finite only. The operators L_γ commute with derivatives and are uniformly bounded in the norm of the space $B^p(\mathbf{R}^n)$. This implies immediately that the L_γ are uniformly bounded in the norm $||\cdot||_{k,p}$ for $k \geqslant 0$. By duality the same is true for $k<0$, due to equality (6.17). Now, by the uniform boundedness of the L_γ it suffices to establish the existence of the desired limit on some dense subset. For example, we may take $\mathrm{CAP}^\infty(\mathbf{R}^n)$ and use Proposition 6.1. □

REMARK 6.4. The space $W^{k,2}(\mathbf{R}_B^n)$ can be described in terms of the Fourier-Bohr transform. To do this we define the space $H^s(\mathbf{R}_B^n)$, $s \in \mathbf{R}$, as the completion of $\mathrm{Trig}(\mathbf{R}^n)$ with respect to the norm

$$||f||_s^2 = \sum_{\xi \in \mathbf{R}^n} (1 + |\xi|^2)^s |\hat{f}(\xi)|^2. \tag{6.18}$$

As usual we put

$$H^\infty(\mathbf{R}_B^n) = \bigcap_{s \in \mathbf{R}} H^s(\mathbf{R}_B^n), \quad H^{-\infty}(\mathbf{R}_B^n) = \bigcup_{s \in \mathbf{R}} H^s(\mathbf{R}_B^n),$$

There are obvious embeddings $H^s(\mathbf{R}_B^n) \subset H^r(\mathbf{R}_B^n)$ for $r \leqslant s$. The spaces $H^s(\mathbf{R}_B^n)$ and $H^{-s}(\mathbf{R}_B^n)$ are dual with respect to the scalar product $(\cdot,\cdot)_B$. A simple computation using (6.15) shows that the norms $||\cdot||_{k,2}$ and $||\cdot||_k$ are equivalent for $k \in \mathbf{Z}$, thus $W^{k,2}(\mathbf{R}_B^n) = H^k(\mathbf{R}_B^n)$. It is easy to observe that all the above statements about the differential operators, Bochner- Fejér operators, etc. hold for the spaces $H^s(\mathbf{R}_B^n)$

with non-integer indices. In particular, a differential operator of the form (6.2) defines a continuous linear map

$$A: H^s(\mathbf{R}_B^n) \to H^{s-m}(\mathbf{R}_B^n),\ s \in \mathbf{R}.$$

Proposition 6.3. is valid for $f \in H^s(\mathbf{R}_B^n), s \in \mathbf{R}$. Note, however, that for $p \neq 2$ the spaces $W^{k,p}(\mathbf{R}_B^n)$ do not coincide with the spaces $H_p^k(\mathbf{R}_B^n)$ introduced in [91]. □

REMARK 6.5. The spaces $W^{s,p}(\mathbf{R}_B^n)$ with non- integer s may be defined by using interpolation theory. This may be done in just the same way as for ordinary Sobolev spaces. □

Comments

A traditional account of the theory of a.p. functions and the history of the classical period of its development may be found in [40, 41, 100]. For a.p. functions defined on general topological groups see [17]. Necessary information on the theory of locally compact abelian groups and on abstract harmonic analysis may be found in the monographs [73, 87]. Some properties of the Bohr compactification are expounded in [85].

§ 1. The account of the theory of Bohr a.p. functions mainly follows [91] and turns out to be a specialization of the more general treating [17] to the case of locally compact abelian groups. The main difference is the consideration of vector-valued functions, which is important in the theory of evolution differential equations. The operators itroduced appear in Bochner's work [113]. More general questions of multiplication by a.p. operator-functions in various spaces of a.p. functions are studied in the author's work [53].

§ 2. The contents of n°. 2.2-2.3 partly agrees with [91], but is more detailed. In contrast to the last paper, the vector case is considered and the spaces B^p of Besicovitch a.p. functions are investigated (in [91] $p=2$). The treatment in terms of almost periods may be found in the classical book [106]. It seems that the description of Besicovitch a.p. functions as a completion of CAP($\mathbf{R}^n$) in a suitable norm (the end of n°. 2.2) goes back to Marcinkiewicz [141]; it has become traditional by now. The investigation of Banach space properties of $\mathcal{B}^p(\mathbf{R}^n)$ is done in [135]. In particular, the non-reflexivity of $\mathcal{B}^p(\mathbf{R}^n)$ is established there. The construction of Bochner-Fejér operators given in the text was suggested in [91] and differs from the traditional one. The material of n°. 2.4 is well-known but is treated here from the point of view of abstract harmonic analysis. As was noted above, the exact description of the spaces $F_B(B^p(G;E))$ is not known yet. However, in the case when $E=H$ is a Hilbert space the classical Hausdorff-Young theorem about Fourier series. may be generalized (together with the proof). Namely: 1) $F_B: B^p(G;H) \to l^{p'}((G')_d;H)$, $1<p\leqslant 2$, is a contraction; 2) if $\phi=(\phi_\xi)\in l^p((G')_d;H)$, $1<p\leqslant 2$, then there is an $f\in B^{p'}(G;H)$ such that $F_Bf=\phi$. A

vast literature is devoted to the harmonic analysis of a.p. functions on **R**; in particular to questions concerning convergence of Fourier-Bohr series $\sum\hat{f}(\xi)\exp(i\xi\cdot t)$. Classical results of this type can be found in [40].

§ 3. The abstract form of the Kronecker-Weyl theorem for scalar functions was obtained in [91]. Here an immediate generalization of that theorem to the case of vector-valued functions is given. Another abstract version of the Kronecker-Weyl theorem may be found in [87]. The notion of relative Bohr compactification in a similar form may be found in [87]. Proposition 3.4 is well-known (even in the case $G=\mathbf{R}^n$). Apparently, Propositions 3.3 and 3.5 were not known in this form earlier.

§ 4. The presentation of the theory of Stepanov a.p. functions slightly differs from the standard one [40, 41].

§ 5. Weakly a.p. functions were introduced by L. Amerio [41, 47, 100]. The contents of n°. 5.1 and 5.2 follows basically the standard presentation. The use of Bohr compactifications (n°. 5.3-5.5) seems to be new in this theory, and, at least in our opinion, clarifies the situation. Theorem 5.5 and Proposition 5.7 did not appear earlier. Theorem 5.8 and Proposition 5.9 are due to Amerio [99]. The criteria for strong almost periodicity were stated by Amerio too, but the proofs (which use the Bohr compactification) are new. An exposition of the theory of weakly a.p. Stepanov functions may be found in [100]. Proposition 5.13 and Theorem 5.14 were obtained in the author's work [67]. Note that a large part of the theory of weakly a.p. functions can be extended to locally compact abelian groups.

§ 6. The discussion follows the papers [89, 91]. The substance of n°. 6.2 has not been published earlier.

CHAPTER 2.

Preliminaries.

1. Some integral inequalities.

Here we derive some estimates which will be used later as technical tools in the investigation of bounded solutions of evolution equations and inequalities, their regularity, almost periodicity, etc. Formerly (see Ch. 1, n°. 4.1) the spaces of Stepanov bounded functions on $\mathbf{R}$ have been introduced. In a similar way Stepanov boundedness on a half line may be defined. More precisely, the space $BS^p(\mathbf{R}_+; E)$, $1 \leqslant p < \infty$, consists of the functions $f \in L^p_{\mathrm{loc}}(\mathbf{R}_+; E)$ for which the form

$$||f||_{S^p} = \sup_{t \in \mathbf{R}_+} \left[\int_t^{t+1} ||f(\tau)||_E^p d\tau \right]^{1/p} \tag{1.1}$$

is finite.

Now let us consider the integral inequality

$$\lambda^2(t)\Big|_{t_1}^{t_2} + \alpha \left[\int_{t_1}^{t_2} \psi^p(t) dt \right]^{q/p} \leqslant \beta \int_{t_1}^{t_2} \psi(t)\theta(t) dt, \tag{1.2}$$

$$\forall t_1, t_2 \in \mathbf{R}_+, \ 0 \leqslant t_2 - t_1 \leqslant 3.$$

Here and below the usual notation $f(t)\big|_{t_1}^{t_2} = f(t_2) - f(t_1)$ is being used. We assume that $\alpha > 0, \beta > 0, p > 1, q > 1$ and that all functions $\lambda \in C(\mathbf{R}_+)$, $\psi \in L^p_{\mathrm{loc}}(\mathbf{R}_+)$, $\theta \in L^{p'}_{\mathrm{loc}}(\mathbf{R}_+)$ are nonnegative.

LEMMA 1.1. *Suppose that $\lambda(t) \leqslant k\psi(t), k > 0$, for almost all $t \in \mathbf{R}_+$, and let $\theta \in BS^{p'}(\mathbf{R}_+)$. Then we have $\lambda \in C_b(\mathbf{R}_+)$ and $\psi \in BS^p(\mathbf{R}_+)$. Moreover, the estimates*

$$\lambda^2(t) \leqslant C \max \left[||\theta||_{S^{p'}}^{q/(q-1)} + k^2 ||\theta||_{S^{p'}}^{2/(q-1)}, ||\lambda^2||_{L^\infty(0,2)} \right], \ t \in \mathbf{R}_+, \tag{1.3}$$

$$||\psi||_{S^p} \leqslant C \max \left[||\theta||_{S^{p'}}^{1/(q-1)} + k^{2/q} ||\theta||_{S^{p'}}^{2/q(q-1)}, ||\lambda^2||_{L^\infty(0,2)} \right] \tag{1.4}$$

are valid, where the constant $C>0$ depends on α, β, p, and q only. If, in addition, $\lambda(0)=0$, then we have the estimates

$$\lambda^2(t) \leqslant C\left[\|\theta\|_{S^{p'}}^{q/(q-1)}+k^2\|\theta\|_{S^{p'}}^{2/(q-1)}\right],\ t\in\mathbf{R}_+, \tag{1.5}$$

$$\|\psi\|_{S^{p}} \leqslant C\left[\|\theta\|_{S^{p'}}^{1/(q-1)}+k^{2/q}\|\theta\|_{S^{p'}}^{2/q(q-1)}\right]. \tag{1.6}$$

Proof. Set

$$\Theta = \|\theta\|_{S^{p'}},\ \psi = \|\psi\|_{S^p},$$

$$\Lambda = \|\lambda^2\|_{L^\infty(\mathbf{R}_+)},\ \Lambda_0 = \|\lambda^2\|_{L^\infty(0,2)}.$$

We will proved that for any integer $n\geqslant 2$ the inequality

$$\lambda^2(t) \leqslant C\max\left[k^2\Theta^{2/(q-1)}+\Theta^{2/(q-1)},\ \Lambda_0\right],\ t\in[0,n], \tag{1.7$_n$}$$

is valid for some $C>0$ not depending on n, λ, ψ, θ, and k. Obviously, this implies inequality (1.3).

Inequality (1.7_2) is trivially valid for $C\geqslant 1$. Now we prove that (1.7_n) implies (1.7_{n+1}), provided the constant C is chosen in a suitable way. Define $\tau_j\in[j-1,j]$ by the equation

$$\lambda(\tau_j) = \max_{t\in[j-1,j]} \lambda(t).$$

If $\lambda(\tau_{n-1})\geqslant\lambda(\tau_{n+1})$, then ($1.7_n$) implies ($1.7_{n+1}$) with the same constant C. Therefore we assume that $\lambda(\tau_{n-1})<\lambda(\tau_{n+1})$. Inequality (1.2) with $t_1=\tau_{n-1}$ and $t_2=\tau_{n+1}$ gives

$$\left[\int_{\tau_{n-1}}^{\tau_{n+1}} \psi^p(t)dt\right]^{(q-1)/p} \leqslant \alpha^{-1}\beta\left[\int_{\tau_{n-1}}^{\tau_{n+1}} \theta^{p'}(t)dt\right]^{1/p'} \leqslant c_0\Theta \tag{1.8}$$

(note that $1\leqslant\tau_{n+1}-\tau_{n-1}\leqslant 3$). Here and below the c_i denote various positive constants not depending on α, β, p, and q only. Using the inequality $\lambda\leqslant k\psi$, we obtain

$$\left[\int_{\tau_{n-1}}^{\tau_{n+1}} \lambda^p(t)dt\right]^{(q-1)/p} \leqslant c_0k^{q-1}\Theta.$$

Hence, there is a $t_0 \in [\tau_n - 1, \tau_n + 1]$ such that $\lambda(t_0) \leqslant c_1 k \Theta^{1/(q-1)}$. Applying (1.2) to the interval $[t_0, \tau_n + 1]$ and using (1.8) we obtain the inequalities

$$\lambda^2(\tau_n+1) \leqslant \lambda^2(t_0) + c_2 \Theta^{q/(q-1)} \leqslant c_3 \left[k^2 \Theta^{2/(q-1)} + \Theta^{2/(q-1)}\right].$$

Therefore, (1.7_{n+1}) is valid, provided $C = c_3$ (which does not depend on $n\,\lambda, \psi, \theta,$ and k).

Now, using (1.2) and (1.3) we obtain

$$\psi^q \leqslant C \cdot (\psi \cdot \Lambda + \Lambda) \leqslant c_1 \max\left[\psi\Theta, \Theta^{q/(q-2)}, k^2 \Theta^{2/(q-1)}, \Lambda_0\right].$$

This implies (1.4).

To prove the second part of the lemma we extend the functions λ, ψ and θ on the interval $[-2, 0]$ by zero. These extensions satisfy all the assumptions of the first part of the lemma for the half line $[-2, +\infty)$. Now, (1.3) and (1.4) imply the estimates (1.5) and (1.6), respectively. The proof is complete. □

REMARK 1.2. Under assumptions of Lemma 1.1 the memberships $\lambda \in C_b(\mathbf{R}_+)$ and $\psi \in BS^p(\mathbf{R}_+)$ are still valid if we replace (1.2) by the more general inequality

$$\lambda^2(t)\,\Big|_{t_1}^{t_2} + \alpha \left[\int_{t_1}^{t_2} \psi^p(t)dt\right]^{q/p} + \alpha_1 \leqslant \beta \int_{t_1}^{t_2} \psi(t)\theta(t)dt, \tag{1.2}$$

$$\forall t_1, t_2 \in \mathbf{R}_+,\ 0 \leqslant t_2 - t_1 \leqslant 3,$$

where $\alpha_1 \in \mathbf{R}$ is an arbitrary constant. Moreover, for the norms we have the following estimates:

$$\|\lambda\|_{L^\infty(\mathbf{R}_+)} \leqslant C_1(\|\theta\|_{S^{p'}}, \|\lambda\|_{L^\infty(0,2)}),$$

$$\|\lambda\|_{S^p} \leqslant C_2(\|\theta\|_{S^{p'}}, \|\lambda\|_{L^\infty(0,2)}),$$

where $C_i(x,y)$, $i = 1,2$, are nonnegative functions which are increasing in each variable (their exact form is not of interest to us). □

We need estimates for solutions of certain other integral inequalities too. Let J be an (open or closed) half-line or $J = \mathbf{R}$.

LEMMA 1.3. *Assume* $\lambda \in C(J), \lambda \geqslant 0$ *and*

$$\lambda^2(t)\big|_{t_1}^{t_2}+\alpha\int_{t_1}^{t_2}\lambda^q(t)dt \leqslant 0,\ t_1, t_2\in J,\ t_1 \leqslant t_2, \tag{1.9}$$

for some $\alpha>0$ and $q\geqslant 2$. Then we have

$$\lambda(t)\underset{(\geqslant)}{\leqslant}\left[\frac{c}{c_1\lambda(t_0)^{q-2}(t-t_0)+c}\right]^{1/(q-2)}\cdot\lambda(t_0) \tag{1.10}$$

for $t\geqslant t_0$ (respectively, for $t\leqslant t_0$), if $q>2$, and

$$\lambda(t)\underset{(\leqslant)}{\geqslant}e^{-c(t-t_0)}\lambda(t_0) \tag{1.11}$$

for $t\geqslant t_0$ (respectively, for $t\leqslant t_0$), if $q=2$. Here $c>0$ and $c_1>0$ depend on α and q only.

Proof. By usual results on integral inequalities [82] (1.9) implies the inequality

$$\lambda^2(t)\underset{(\geqslant)}{\leqslant}y(t),\ t \geqslant t_0\,(t \leqslant t_0),$$

where $y(t)$ is the solution of the differential equation $y'=-\alpha y^{q/2}$ with initial value $y(t_0)=\lambda^2(t_0)$. By solving this equation we obtain the required estimates. □

In the proof of the following lemma we need the Hölder inequality with exponents lying outside the interval $[1, +\infty]$ [2], i.e.

$$\int\psi\theta dt \geqslant \left(\int\psi^r dt\right)^{1/r}\left(\int\theta^{r'} dt\right)^{1/r'},\ \frac{1}{r}+\frac{1}{r'} = 1, \tag{1.12}$$

where $\psi(t)\geqslant 0$, $\theta(t)>0$, $0<r<1$.

LEMMA 1.4. *Suppose that $s\in BS^p(J)$, $s>0$ and $\|s\|_{S^p}\leqslant R$. Assume that $\lambda\in C(J)$, $\lambda\geqslant 0$ satisfies the inequality*

$$\lambda^2(t)\big|_{t_1}^{t_2}+\alpha\int_{t_1}^{t_2}\lambda^q(t)s^{p-q}(t)dt \leqslant 0,\ t_1, t_2\in J,\ t_1 \leqslant t_2, \tag{1.13}$$

where $q>p>1$ and $\alpha>0$. Then for any $t_0\in J$ and $\delta>0$ the estimates

$$\lambda(t) \leqslant c\frac{\lambda(t_0)^{2/q}}{(t-t_0)^{1/q}},\ t \geqslant t_0+\delta, \tag{1.14}$$

$$\lambda(t) \geqslant c_1(t_0-t)^{1/2}\lambda(t_0),\ t \leqslant t_0-\delta, \tag{1.15}$$

are valid, where $c>0$ and $c_1>0$ depend on α, p, q, R, and δ only.

Proof. Using (1.12) with $r=p/q$ and $r=p/(p-q)$, we deduce from (1.13) the inequality

$$\lambda^2(\tau)\,|_{t_1}^{t}+\alpha\left[\int_{t_0}^{t}\lambda^p(\tau)d\tau\right]^{1/p}\left[\int_{t_0}^{t}s^p(\tau)d\tau\right]^{(p-q)/p}\leqslant 0.$$

For $t\geqslant t_0+\delta$ the second integral does not exceed

$$\alpha_1(t-t_0)\cdot\sup_{0\leqslant b-a\leqslant 1}\int_a^b s^p(\tau)d\tau\leqslant\alpha_1R^p(t-t_0),$$

where $\alpha_1>0$ depends on δ only. Since $p<q$, we obtain the inequality

$$\lambda^2(\tau)\,|_{t_0}^{t}+\alpha_2(t-t_0)^{(p-q)/p}\left[\int_{t_0}^{t}\lambda^p(\tau)d\tau\right]^{q/p}\leqslant 0,\ t\geqslant t_0+\delta. \tag{1.16}$$

As an immediate consequence of (1.13) we claim that λ is monotone decreasing. Therefore, replacing in (1.16) $\lambda^p(\tau)$ by $\lambda^p(t)$, we have the inequality

$$\lambda^2(t)-\lambda^2(t_0)+\alpha_2(t-t_0)\lambda^q(t)\leqslant 0,\ t\geqslant t_0+\delta.$$

Forgetting the summand $\lambda^2(t)$, we obtain (1.14). Inequality (1.15) is an immediate consequence of (1.14) (interchange t and t_0). □

2. Composition operators.

2.1. COMPOSITION OPERATORS IN L^p-SPACES

Let X be a locally compact topological space with measure* μ, let E and F be Banach spaces, and let $A(x):E\to F$, $x\in X$, be a family of (nonlinear) operators, satisfying the

Carathéodory condition: For μ-almost all $x\in X$ the map $A(x):E\to F$ is continuous and for any $v\in E$ the F-valued function $A(x)v$ is measurable in $x\in X$.

Now we consider the operator $\mathbf{A}:f(x)\mapsto A(x)f(x)$.

PROPOSITION 2.1. *Assume μ is non-atomic, i.e. for any μ-measurable subset $S\subset X$ there is a μ-measurable subset $S_1\subset S$ with the property $\mu(S_1)=\mu(S)/2$. Suppose that*

* Standard references are [4, 5].

the Carathéodory condition is satisfied and that for μ*- almost all* $x\in X$ *the inequality*

$$||A(x)v||_F \leqslant a(x)+b||v||_E^{p_1/p_2},\ v\in E, \tag{2.1}$$

is valid with $a\in L^{p_2}(X)$. *Then the operator* $\mathbf{A}$ *maps* $L^{p_1}(X;E)$ *into* $L^{p_2}(X;F)$, *and it is bounded and continuous.*

The *proof* in case E and F are both finite dimensional, X is an open subset in $\mathbf{R}^n$ and m is Lebesgue measure can be found in [32]. The general case is considered in [34]. □

Recall that a nonlinear operator is bounded if it transforms any bounded set to a bounded set.

There is a case of interest in which the Carathéodory condition is not satisfied. Let $E=V$ be a real reflexive Banach space and let $F=V'$ be its dual. Recall that an operator $A:V\to V'$ is said to be

(a) monotone if

$$(Av-Aw, v-w) \geqslant 0\ \forall v,\ w\in V; \tag{2.2}$$

(b) semicontinuous if for any $u, v, w\in V$ the $\mathbf{R}$-valued function $\lambda\mapsto(A(u+\lambda v),w)$ is continuous.

(See [44].)

Instead of Carathéodory condition we introduce the following condition:

(c) for any $x\in X$ *the operator* $A(x):V\to V'$ *is monotone, bounded and semicontinuous; for any* $v\in V$ *and any bounded subset* $U\subset V$ *the family* $\{(A(x)u, v)\,|\,u\in U\}$ *is equicontinuous on any compact subset of* X.

PROPOSITION 2.2. *Assume that* X *is a locally compact metrizable space, that condition* (c) *is valid and that* $A(x)$ *satisfies the inequality*

$$||A(x)v||_* \leqslant c||v||^{p-1}+a(x),\ v\in V,$$

where $p>1, c>0$ *are constants independent of* $x\in X$ *and* $a\in L^{p'}(X)$. *Then* $\mathbf{A}$ *is a monotone semicontinuous operator, acting from* $L^p(X;V)$ *into* $L^{p'}(X;V')$, *and*

$$||\mathbf{A}v||_{L^{p'}(X;V')} \leqslant c_1||v||^{p-1}_{L^p(X;V)}+c_2,\ v\in L^p(X;V) \tag{2.3}$$

(in particular, $\mathbf{A}$ *is bounded).*

Proof. We prove that for any measurable V-valued function $v(x)$ the V'-valued function $A(x)v(x)$ is also measurable. First, its weak measurability (i.e. the measurability of all functions $(A(x)v(x), w), w\in V)$ will be established. As $v(x)$ is measurable, for any compact subset $K\subset X$ there are a subset $N\subset K$ having measure zero and a collection $\{K_n\}, n\in\mathbf{N}$, of compact sets such that $K\setminus N=\bigcup K_n$ and all restrictions $V|_{K_n}$ are continuous functions. Now we claim that $(A(x)v(x), w)$ is continuous on K_n. Indeed, suppose that $x=\lim x_k$, $x_k \in K_n$.

It is obvious that

$$|(A(x)v(x)-A(x_k)v(x_k), w)| \leqslant$$

$$\leqslant |(A(x)v(x)A(x)v(x_k), w)|+|(A(x)v(x_k)-A(x_k)v(x_k), w)|.$$

Here the right-hand side tends to zero. For the first term at the right this is true because $A(x)$ is a continuous operator from the space V endowed with the strong topology into the space V endowed with the weak topology [44, 118]. For the second term the required follows by condition (c). This implies the weak measurability.

Now to prove the measurability of $A(x)v(x)$ we need to show that the function $A(x)v(x)|_K$ is almost separably-valued. Since X is measurable, any K_n contains a dense countable subset. Let $\{x_i\}$ be in the union of those subsets and let W be the closed linear hull of all vectors $A(x_i)v(x_i)$. For any $x \in K \setminus N$ there is a sequence $\{x_{i'}\}$ such that $x=\lim x_{i'}$. So $A(x_{i'})v(x_{i'}) \to A(x)v(x)$ weakly in V'. Since W is weakly closed, we have $A(x)v(x) \in W$. Hence $A(x)v(x)$ is measurable.

The inequality (2.3) and mononicity of **A** follow from corresponding inequalities for $A(x)$ by integration.

To prove the semicontinuity of **A** we consider a function $<\mathbf{A}(u+\lambda v), w>$, where $u, v, w \in L^p(X; V)$, and we show its continuity at the point $\lambda=0$. Here $<\cdot,\cdot>$ denotes the pairing between $L^p(X; V)$ and $L^{p'}(X; V')$. Since the set of μ-simple functions is dense in $L^p(X; V)$, and in view of (2.3), we may suppose without loss of generality that $w=\chi \cdot w_0$, where $w_0 \in V$ and χ is the characteristic function of a measurable subset $M \subset X$ with $\mu(M)<\infty$. By (2.3), for any $\epsilon>0$ there is a compact subset $K \subset X$ such that the functions $u|_K$ and $v|_K$ are continuous and

$$\int_{X\setminus K} ||\mathbf{A}(u+\lambda v)||_*^{p'} d\mu(x) < \epsilon, \quad |\lambda| \leqslant 1.$$

Choosing an appropriate δ-net in K and using (c) we may approximate $\mathbf{A}(u+\lambda v)$ in the space $L^{p'}(K; V')$ by a function $F(x, \lambda)$ of the form $F(x,\lambda)=A(x_i)(u(x)+\lambda v(x))$, $x \in U_i$, where $\bigcup_{i=1}^m U_i=K$ up to a set of zero measure and with $x_i \in U_i$. Such a function $F(x, \lambda)$ is measurable on $K \times [-1, 1]$ and for $x \in \bigcup U_i$ it is weakly continuous in λ. By Lebesgue's theorem,

$$\int_{U_i \cap M} (F(x, \lambda), w_0) d\mu(x)$$

depends continuously on λ for $|\lambda|\leqslant 1$. Since

$$|\langle \mathbf{A}(u+\lambda v), w\rangle - \int_{K\cap M} (F(x,\lambda), w_0)d\mu(x)| \leqslant C\epsilon,$$

where $C>0$ depends on $\mathbf{A}$, u, v, and w only, and $\epsilon>0$ may be taken arbitrarily small, the semicontinuity of $\mathbf{A}$ is proved. □

REMARK 2.3. Under the conditions of Proposition 2.2, assume that the operators $A(x)$ are uniformly coercive in the sense that

$$(A(x)v, v) \geqslant \alpha||v||^p+\beta(x),\ v\in V, \tag{2.4}$$

where $\alpha>0$ and $\beta\in L^1(X)$. Then the operator $\mathbf{A}$ is also coercive:

$$\langle \mathbf{A}v, v\rangle \geqslant \alpha||v||^p_{L^p(X;V)}+\gamma,\ v\in L^p(X;V), \tag{2.5}$$

with $\gamma=\int_X \beta(x)d\mu(x)$.

2.2. COMPOSITION OPERATORS IN BESICOVITCH SPACES.

Let G be a locally compact abelian group and $A(x): V\to V'(x\in G)$ a family of monotone bounded semicontinuous operators. As before, we assume that V is a real reflexive Banach space. The following condition is assumed to hold.

(cap) *for any bounded $U\subset V$ all functions $\{A(x)u\,|\,u\in U\}$ with values in V' are equicontinuous and uniformly almost periodic.*

PROPOSITION 2.4. *There is a family $\tilde{A}(x): V\to V'$ $(x\in G_B)$ of monotone bounded semicontinuous operators such that for any bounded $U\subset V$ the functions $\tilde{A}(x)u$, $u\in U$, are equicontinuous on G_B and $\tilde{A}(x)=A(x)$ for $x\in G$. Moreover, A is of the form $\tilde{A}(x)=\tilde{A}_S(p_s x)$, where $S\subset(G')_d$ is a countable subgroup, $p_S: G_B\to G_{B,S}$ is the canonical epimorphism and $\tilde{A}_S(y): V\to V'$ $(y\in G_{B,S})$ is a family of monotone bounded semicontinuous operators such that the functions $\tilde{A}_S(y)u$, $u\in U$, are equicontinuous on $G_{B,S}$.*

Proof. Let Y be the space of all bounded maps from V into V'. Set

$$\rho_N(A_1, A_2) = \sup_{||v||\leqslant N} ||A_1 v - A_2 v||_*,$$

$$\rho(A_1, A_2) = \sum_{N=1}^{\infty} 2^{-N}\frac{\rho_N(A_1, A_2)}{1+\rho_N(A_1, A_2)},$$

for $A_1, A_2\in Y$. The set Y endowed with the metric ρ is a complete metric space.

The metric ρ defines the topology of uniform convergence on bounded sets. The family A defines an a.p. function $A:G\to Y$. To construct $\tilde{A}_S$ we apply Proposition 3.5 of Ch. 1. Since the function $\tilde{A}_S:G_{B,S}\to Y$ is continuous, boundedness, monotonicity and semicontinuity of the operators $\tilde{A}_\rho(y)$ follow from the fact that these hold on the dense subset $G\subset G_{B,S}$ of y's. □

The minimal subgroup $S\subset(G')_d$ having the property described in Proposition 2.3 will be denoted by $S(A)$ and will be called the *frequency group* of the family A.

Now we define the composition operator $\mathbf{A}$ by $(\mathbf{A}v)(x)=A(x)v(x)$.

PROPOSITION 2.5. *Assume that condition (cap) is valid and that*

$$||A(x)v||_* \leqslant C||v||^{p-1}+c_1,\ v\in V, \tag{2.6}$$

where $p>1$, $c>0$, and $c_1\in\mathbf{R}$ does not depend on $x\in G$. Then $\mathbf{A}$ is a monotone semicontinuous operator from $B^p(G;V)$ into $B^{p'}(G;V')$, and

$$||\mathbf{A}v||_{B^{p'}} \leqslant C||v||_{B^p}^{p-1}+C_1,\ v\in B^p(\mathbf{R};V), \tag{2.7}$$

where $C>0$.

Proof. We consider the family of operators $\tilde{A}_{S_0}$, where $S_0=S(A)$. A bound similar to (2.6) is valid for $\tilde{A}_{S_0}$ by continuity. Let $S\subset(G')_d$ be a countable subgroup such that $S_0\subset S$. For a V-valued function $v(x)$ on $G_{B,S}$ we set

$$(A_Sv)(x) = \tilde{A}_{S_0}(p_{S_0,S}x)v(x),$$

where $p_{S_0,S}:G_{B,S}\to G_{B,S_0}$ is the canonical epimorphism. The group $G_{B,S}$ is metrizable (see Ch. 1, Remark 3.6) and, as a consequence, Proposition 2.2 is applicable. Hence $\mathbf{A}_S:L^p(G_{B,S};V)\to L^{p'}(G_{B,S};V')$ is a monotone semicontinuous operator for which a bound of type (2.3) holds. Under the identification $p_S^*L^r(G_{B,S};E)\simeq B_S^r(G;E)$ (see Ch. 1, Proposition 3.3) the operator $\mathbf{A}_S$ turns out to be the restriction of $\mathbf{A}$ to the subspace $B_S^p(G;V)$. Combining this fact with Proposition 3.4 of Ch. 1 we obtain the required assertion. □

REMARK 2.6. Next to the condition (cap), below we will also use the following more restrictive assumption ($p>0$):

(cap$_{V,p}$) *the family $\{(1+||v||^{p-1})^{-1}A(x)v \mid v\in V\}$ of V'-valued functions is equicontinuous and uniformly a.p.*

In the space $Y_{p,V}$ of all operators from V into V' satisfying the inequality (2.6) (with constants depending on the operator) we introduce the metric

$$d_{p,V}(A_1,A_2) = \sup_{v\in V}\frac{||A_1v-A_2v||_*}{1+||v||^{p-1}}. \tag{2.8}$$

The metric space $Y_{p,V}$ is complete and condition $(\mathrm{cap}_{V,p})$ is equivalent to almost periodicity of the $Y_{p,V}$-valued function $A(x)$.

REMARK 2.7. Let E and F be Banach space. Consider a family of operators $A(x):E\to F$ satisfying the evident extension of the condition (cap). If all operators $A(x), x\in G$, are bounded and continuous, then the same is true for all operators $\tilde{A}(y), y\in G_B$ (obviously, the corresponding assertion of Proposition 2.4 is valid in the case under consideration). Moreover, if the group G is infinite, then the Haar measure μ on G_B is non-atomic. Indeed, G_B is infinite too. Now if $\mu(\{x_0\})>0$, then the invariance property of μ implies $\mu(\{x\})=\mu(\{x_0\})>0$ for any $x\in G_B$. Because of the property $\mu(G_B)=1$, we obtain a contradiction. Hence, μ is non-atomic and Proposition 2.1 is applicable. So the inequality

$$||A(x)v||_F \leqslant a+b||v||_E^{p_1/p_2}, \quad v\in E \tag{2.9}$$

with $a\in\mathbf{R}$ and $b>0$ implies that the operator $\mathbf{A}:B^{p_1}(G;E)\to B^{p_2}(G;F)$ is bounded and continuous for any infinite group G (for finite groups this is trivial).

2.3. SOMETHING ON CONVEX FUNCTIONS

(See, for example, [96].) Let ϕ be a convex function on a real reflexive Banach space V with domain $D(\phi)=\{v\in V\,|\,\phi(v)<\infty\}$. Recall that such a function ϕ is said to be

a) lower semicontinuous (or closed) if its epigraph $E(\phi)=\{(v,\xi)\in V\times\mathbf{R}\,|\,v\in D(\phi), \xi\geqslant\phi(v)\}$ is closed.

b) proper if $\phi\not\equiv+\infty$, i.e, $D(\phi)\neq\varnothing$.

For the general theory of convex functions see, e.g., [96]

PROPOSITION 2.8. *Let X be a compact topological space endowed with a finite measure μ and let ϕ be a proper lower semicontinuous convex function on a reflexive space V. Then for any $v\in L^1(X;V)$ the function $\phi(v)$ is measurable on X. It is integrable iff $\phi(v(x))\leqslant g(x)$ almost everywhere for some function $g\in L^1(x)$. For each $p\in[1,\infty]$,*

$$\Phi(v)=\begin{cases}\int_X \phi(v(x))d\mu(x), & \phi(v)\in L^1(X),\\ +\infty, & \phi(v)\notin L^1(X),\end{cases}$$

is a proper lower semicontinuous convex function on $L^p(X;V)$.

Proof. See [118, Appendix I]. □

REMARK 2.9. Obviously, the previous proposition is applicable in the case when

$X=G_B$ and μ is the Haar measure. So

$$\Phi(v) = \Phi_B(v) = \begin{cases} \mathbf{M}\{\phi(v)\}, & \phi(v)\in B^1(G), \\ +\infty, & \phi(v)\notin B^1(G), \end{cases}$$

is a proper semicontinuous convex function on $B^p(G; V)$, $1\leqslant p<+\infty$.

Comments

§ 1. The simplest assertion of the type of Lemma 1.1 was given in implicit form in the paper [144] of G. Prouse, where it was used to prove the boundedness of solutions of abstract nonlinear parabolic equations on a half-line and to obtain existence theorems for bounded solutions on a line. Similar arguments were subsequently used by many authors [22, 25, 41, 44, 100, 109, 110]. In the case when $p=q$ and inequality (1.2) is satisfied for all $t_1<t_2$, Lemma 1.1 without estimates (1.3)-(1.6) was stated explicitly in [110]. In general form this statement was established by the author [57, 67]. These papers (see also Ch. 3 and 4) contain some new applications of such estimates (to prove, in particular, time regularity of bounded solutions). Note that in many cases, but not always, inequality (1.2) is fulfilled for all $t_1<t_2$. Of course, the restriction $0\leqslant t_2-t_1\leqslant 3$ may be replaced by $0\leqslant t_2-t_1\leqslant\text{const}$. It seems that in this explicit form Lemma 1.4 has not appeared earlier.

§ 2. A large number of papers deals with composition operators under the Carathéodory condition (see, for example, [6, 32]). Proposition 2.2 is well-known, but the author has not found a proof in the literature. The case when $A(x)\equiv A$ is considered in [118] and the case when $X=[a, b]$ and μ is Lebesgue measure is considered in [57]. In a slightly less general form Proposition 2.4 is contained implicitly in the author's papers [60, 67].

CHAPTER 3.

Solutions of evolution variational inequalities bounded and almost periodic in time.

1. On variational inequalities.

Here we collect some preliminaries on evolution inequalities of the first order in time. Detailed accounts of the theory and various applications can be found in [21, 44, 118].

1.1. STATEMENT OF THE PROBLEM AND WEAK SOLUTIONS.

Let V be a real reflexive Banach space and let H be a Hilbert space, identified with its dual H'. We assume that $V \subset H \subset V'$ and that all inclusions are dense and continuous. By $(\cdot, \cdot)$ we denote the inner product in H and the canonical bilinear form on $V' \times V$. The symbols $\|\cdot\|$, $|\cdot|$ and $\|\cdot\|_*$ stand for the norms in V, H and V', respectively.

Now let $A(t): V \to V$, $t \in J$, be a family of operators defined on some interval J (below $J = \mathbf{R}$ as a rule) and let ϕ be a convex function on V. Given a function $f: J \to V'$ we consider the following problem (see, for example, [44]). Find a function $u: J \to V$ satisfying the inequality

$$(u'(t) + A(t)u(t) - f(t), v - u(t)) + \phi(v) - \phi(u(t)) \geqslant 0, \ \forall v \in V, \tag{1.1}$$

for almost all $t \in J$. For certain reasons this formulation is not satisfactory and a notion of weak solution of (1.1) is required [44].

So now we will reproduce formal arguments giving rise to a weak formulation of the problem. Taking $v = v(t)$ in (1.1) and integrating over $[t_1, t_2] \subset J$ we obtain

$$\int_{t_1}^{t_2} [(u'(t) + A(t)u(t) - f(t), v(t) - u(t)) + \phi(v(t)) - \phi(u(t))]dt \geqslant 0.$$

So,

$$\int_{t_1}^{t_2}[(v'(t)+A(t)u(t)-f(t),\, v(t)-u(t))+\phi(v(t))-\phi(u(t))]dt =$$

$$= \int_{t_1}^{t_2}[(u'(t)+A(t)u(t)-f(t),\, v(t)-u(t))+\phi((v(t))-\phi(u(t))]dt+$$

$$+\int_{t_1}^{t_2}(v'(t)-u'(t),\, v(t)-u(t))dt \geqslant$$

$$\geqslant \frac{1}{2}\int_{t_1}^{t_2}\frac{d}{dt}\mid v(t)-u(t)\mid^2 dt = \frac{1}{2}\mid v(t)-u(t)\mid^2\mid_{t_1}^{t_2}.$$

Hence (1.1) implies

$$\int_{t_1}^{t_2}[(v'(t)+A(t)u(t)-f(t),\, v(t)-u(t))+\phi(v(t))-\phi(u(t))]dt \geqslant \tag{1.2}$$

$$\geqslant \frac{1}{2}\mid v(t)-u(t)\mid^2\mid_{t_1}^{t_2}\forall t_1,\, t_2\in J,\; t_1 \leqslant t_2,$$

for any test function $v(t)$. If an explicit indication of $A(t)$ and $f(t)$ will be needed, we will refer to (1.2) as $VI(A,f)$.

Now we will fix our basic assumption and describe a precise formulation of the problem.

1. *Operators* A. It is assumed that $A(t)\colon V\to V'$ is a family of monotone semicontinuous operators satisfying the inequality

$$\mid\mid A(t)v\mid\mid_* \leqslant c_1\mid\mid v\mid\mid^{p-1}+c_2,\; v\in V, \tag{1.3}$$

where $p>1$, $c_1>0$ and $c_2\in\mathbf{R}$ are independent of $t\in J$. The following condition (see Ch. 2, § 2) is assumed to be valid:

(c) *for any* $v\in V$ *and any bounded set* $U\subset V$ *the family of functions* $\{(A(t)u,\, v)\mid u\in U\}$ *is equicontinuous on any compact subinterval of* J.

2. *The function* ϕ is assumed to be a lower semicontinuous convex function on V with domain $D(\phi)$.

3. *Test functions*. This name is used for functions $v \in L^p_{loc}(J; V)$ such that $v' \in L^{p'}_{loc}(J; V')$ (the derivative is taken in the weak sense). Any such function lies in $C(J; H)$ (see, for example, [44, 118]).

4. *Weak solutions*. Given $f \in L^{p'}_{loc}(J; V')$, a function $u \in L^p_{loc}(J; V) \cap C(J; H)$ is said to be a weak solution of (1.1), if inequality (1.2) is satisfied for any test function v. For such a solution we have $\phi(u) \in L^1_{loc}(J)$. Note that the derivative u' (in the sense of distributions) of a weak solution need not be of class $L^{p'}_{loc}(J; V')$. Below we will use the term "solution" instead of "weak solution".

5. *Convention*. These notations and assumptions are assumed to be valid in all of this Chapter, unless the converse is stated explicitly.

By Proposition 2.2 of Ch. 2, $\mathbf{A}: u(\cdot) \mapsto A(\cdot)u(\cdot)$ is a bounded monotone semicontinuous operator from $L^p(a, b; V)$ into $L^{p'}(a, b; V')$, $[a, b] \subset J$. Further, by Proposition 2.8 of Ch. 2,

$$\Phi(v) = \begin{cases} \int_a^b \phi(v(t))dt, & \phi(v) \in L^1(a, b), \\ +\infty, & \phi(v) \notin L^1(a, b), \end{cases} \tag{1.4}$$

is a proper lower semicontinuous convex function on $L^p(a, b; V)$. Hence we may apply results of H. Brezis [118]. As a consequence we obtain

PROPOSITION 1.1. *Assume that the family of operators $A(t)$, $t \in \mathbf{R}_+$, satisfies the inequality*

$$(A(t)v, v) \geqslant \alpha \| v \|^p + \beta, \quad v \in V, \tag{1.5}$$

where $\alpha > 0$ and $\beta \in \mathbf{R}$ are independent on $t \in \mathbf{R}_+$. Then for any $f \in L^{p'}_{loc}(\mathbf{R}_+; V')$ and any $u_0 \in \overline{D(\phi)}^H$ (closure in H) there exists a unique solution $u \in L^p_{loc}(\mathbf{R}_+; V) \cap C(\mathbf{R}_+; H)$ of inequality (1.2) such that $u(0) = u_0$. Moreover, we have

$$\| u \|_{L^p(0,T;V)} \leqslant C_1(\| f \|_{L^{p'}(0,T;V')}, | u_0 |),$$

$$\| u \|_{C([0,T];H)} \leqslant C_2(\| f \|_{L^{p'}(0,T;V')}, | u_0 |),$$

where $C_i(x, y)$, $i = 1,2$, $x, y \in \mathbf{R}_+$, *are continuous nondecreasing functions with respect to each variable.*

In [118] it is assumed that the norms $||\cdot||$ and $||\cdot||_*$ are strictly convex. In [102, 120] it is stated that any reflexive Banach space has an equivalent norm, which is strictly convex together with its dual. Recall that the spaces $L^r_{loc}(0, +\infty; \cdot)$ (open half-line) and $L^r_{loc}(\mathbf{R}_+; \cdot)$ (closed half-line) do not coincide.

REMARK 1.2. In case $\phi \equiv 0$ problem (1.1) is equivalent to the equation

$$\frac{du}{dt} + A(t)u(t) = f(t). \tag{1.6}$$

Under our assumptions, let u be a solution of (1.2) with $\phi \equiv 0$. Then $u' \in L^{p'}_{loc}(J; V')$ and u satisfies equation (1.6). Indeed, let $[a, b] \subset J$ be a subinterval. Standard existence theory for equation (1.6) (see, for example, [44]) implies that there is a unique $\tilde{u} \in L^p(a, b; V)$ such that $\tilde{u}' \in L^{p'}(a, b; V')$, $\tilde{u}$ satisfies (1.6) and $\tilde{u}(a) = u(a)$ (of course, $\tilde{u} \in C([a, b]; H)$). Then $\tilde{u}$ is a solution of (1.2) on $[a, b]$. So, by the uniqueness of the solution of the Cauchy problem for inequality (1.2), $\tilde{u}$ coincides with $u|_{[a,b]}$. In particular, Proposition 1.1 gives a solution of the Cauchy problem for equation (1.6)

REMARK 1.3. The following generalization of our setting is very useful [44, 118]. Let V_i, $i = 1, \ldots, n$, be reflexive Banach spaces, $V_i \subset H$, and suppose the embeddings are continuous and dense. Then $V_i \subset H \subset V_i'$. Consider the space

$$V' = \bigcap_{i=1}^{n} V_i,$$

endowed with the usual intersection norm $||u|| = \max_{1 \leq i \leq n} ||u||_i$. Then

$$V' = \sum_{i=1}^{n} V_i',$$

where the norm is defined by the formula

$$||f||_* = \inf\{\sum ||f_i||_{*i} \,|\, f = \sum f_i,\ f_i \in V_i',\ i = 1, \ldots, n\}.$$

Suppose we are given monotone semicontinuous operators $A_i(t): V_i \to V_i'$, $t \in J$, $i = 1, \ldots, n$, satisfying condition (c) and for some $p_i > 1$ inequalities

$$||A_i(t)v||_{*i} \leqslant c_1||v||_i^{p_i-1}+c_2,\ v\in V_i,$$

$$(A_i(t)v, v) \geqslant \alpha||v||_i^{p_i}+\beta,\ v\in V_i,$$

where $c_1>0$, $\alpha>0$, $c_2\in\mathbf{R}$, $\beta\in\mathbf{R}$ are independent on $t\in J$. We define the operator $A(t): V\to V'$ by the formula

$$A(t)v = \sum_{i=1}^{n} A_i(t)v,\ v\in V.$$

Then the following problem makes sense. For a given

$$f\in\sum L_{\mathrm{loc}}^{p_i'}(J; V_i')$$

it is required to find a function

$$u\in[\bigcap L_{\mathrm{loc}}^{p_i}(J; V_i)]\bigcap C(J; H)$$

satisfying inequality (1.2) for any function v for which

$$v\in\bigcap L_{\mathrm{loc}}^{p_i}(J; V_i),\ v'\in\sum L_{\mathrm{loc}}^{p_i'}(J; V_i').$$

In this case there is a statement similar to Proposition 1.1.

1.2. ESTIMATES FOR SOLUTIONS.

Without loss of generality we may assume that in out setting the following conditions are valid:

1) ϕ is nonnegative:
2) $\phi(0)=0$;
3) $A(t)0=0$.

Indeed, this may be achieved by the following series of equivalent transformations of problem (1.2). Let $v_0\in D(\phi)$. Replace u by $u-v_0$. Then the function $\phi(\cdot)$ and the operators $u\mapsto A(t)u$ must be replaced by $\phi(\cdot+v_0)$ and $u\mapsto A(t)(u+v_0)$, respectively. So we may assume that $0\in D(\phi)$. Further, there is an element $\xi\in V'$ such that

$$\phi(v) \geqslant (\xi, v)+\phi(0),\ v\in V,$$

i.e. the support functional to ϕ at zero (see, for example, [96]). It is easy to see that adding to ϕ a constant does not change inequality (1.2). Hence, by subtracting $\phi(0)$ we arrive at the case when $\phi(v) \geqslant (\xi, v)$, $v \in V$, and condition 2) is fulfilled. Now replacing $\phi(v)$ and $f(t)$ by $\phi(v)-(\xi, v)$ and $f(t)-\xi$, respectively, we obtain the validity of 1). Finally, we may consider the operator $u \mapsto A(t)u - A(t)0$ and the function $f(t)-A(t)0$ instead of $A(t)u$ and $f(t)$, respectively, to arrive at the validity of 3).

It is important that all the results obtained below are invariant under the above mentioned reduction. Indeed, the function $A(t)0$ is V'-bounded and V'-measurable (see Ch. 2, § 2), i.e. $A(t)0 \in L^{\infty}(\mathbf{R}; V')$. In particular, adding to f this function or a constant ξ does not violate the condition $f \in BS^{p'}(\mathbf{R}; V')$ in the existence theorems for bounded solutions which we obtain in §§ 2 and 4. There is a similar correspondence between properties of A and f in all our results.

Below conditions 1) - 3) are assumed to be valid. Moreover, we assume, for simplicity, $\beta=0$ in (1.5). Our final results do not depend on this assumption.

LEMMA 1.4. *Let u be a solution of inequality (1.2) on an interval J and assume estimate (1.5) to be valid. Then for any $t_j \in J$, $j=1, 2$, $t_1 \leqslant t_2$, we have*

$$\frac{1}{2} | u(t) |^2 |_{t_1}^{t_2} + \alpha \int_{t_1}^{t_2} \| u(t) \|^p dt + \int_{t_1}^{t_2} \phi(u(t)) dt \leqslant \int_{t_1}^{t_2} (f(t), u(t)) dt. \tag{1.7}$$

Proof. Put $v \equiv 0$ in (1.2) and use (1.5) and condition 2). □

Now we consider functions $u_1, u_2 \in L^p_{loc}(J; V) \cap C(J; H)$ satisfying for any test function v the inequality

$$\int_{t_1}^{t_2} [(v' - g, v-u) + \phi(v) - \phi(u)] dt \geqslant \frac{1}{2} | v-v |^2 |_{t_1}^{t_2} \forall t_1, t_2 \in J, \tag{1.8}$$

with $g=g_1$ and $g_2 \in L^{p'}_{loc}(J; V')$, respectively. In [118, Theorem II.3] it is stated that the following estimate holds:

$$\frac{1}{2} | u_1 - u_2 |^2 |_{t_1}^{t_2} \leqslant \int_{t_1}^{t_2} (g_1 - g_2, u_1 - u_2) dt, \; t_1, t_2 \in J, \; t_1 \leqslant t_2. \tag{1.9}$$

LEMMA 1.5. Let u_i be solutions of *(1.2) with $f=f_i \in L^{p'}_{loc}(J; V')$, $i=1, 2$. Then for any $t_1, t_2 \in J$, $t_1 \leqslant t_2$, the following estimates are valid:*

$$\frac{1}{2}|u_1-u_2|^2|_{t_1}^{t_2}+\int_{t_1}^{t_2}(Au_1-Au_2, u_1-u_2)dt \leqslant \int_{t_1}^{t_2}(f_1-f_2, u_1-u_2)dt, \qquad (1.10)$$

$$\frac{1}{2}|u_1-u_2|^2|_{t_1}^{t_2} \leqslant \int_{t_1}^{t_2}(f_1-f_2, u_1-u_2)dt. \qquad (1.11)$$

Proof. By Proposition 2.2 of Ch. 2 we have $A(\cdot)u_i(\cdot)\in L^{p'}_{loc}(J; V')$, $i=1, 2$, and so $g_i(\cdot)=f_i(\cdot)-A(\cdot)u_i(\cdot)\in L^{p'}_{loc}(J; V')$. For such g_i the functions u_i satisfy inequality (1.8) and (1.9) implies (1.10). By the monotonicity of A inequality (1.11) follows from (1.10). □

1.3. PASSAGE TO THE LIMIT IN VARIATIONAL INEQUALITIES.

The following statement gives one of the possible ways to justify passage to the limit in variational inequalities. Other versions may be found in, for example, [44]. Here we take $J=[T_1, T_2]$.

PROPOSITION 1.6. *Let $u_n\in L^p(J; V)\cap C(J; H)$ be solutions of* (1.2) *with right-hand sides $f_n\in L^{p'}(J; V')$. Assume that $\lim f_n=f$ in $L^{p'}(J; V')$, $\lim u_n=u$ in $C(J; H)$ and weakly in $L^p(J; V)$. Then u is a solution of (1.2) with right-hand side f.*

Proof. Under our assumptions inequality (1.7) implies

$$||\phi(u_n)||_{L^1(J)} \leqslant c, \qquad (1.12)$$

where c is independent of n. The function Φ defined by (1.4) is lower semicontinuous on $L^p(a, b; V)$, $[a, b]\subset J$. Due to convexity this is true both in the strong and in the weak topology. So, by (1.12) we have

$$\int_a^b \phi(u)dt \leqslant \liminf_{n\to\infty} \int_a^b \phi(u_n)dt \leqslant c. \qquad (1.13)$$

In particular, $\phi(u)\in L^1(J)$.

For any test function v and any $a, b\in J$, $a\leqslant b$, we have

$$\lim\frac{1}{2}|v-u_n|^2|_a^b = \frac{1}{2}|v-u|^2|_a^b,$$

$$\lim \int_a^b (v' - f_n, v - u_n)dt = \int_a^b (v' - f, v - u)dt.$$

Hence, to prove our statement we only need obtain the inequality

$$\liminf \int_a^b (Au_n, v - u_n)dt \leq \int_a^b (Au, v - u)dt. \tag{1.14}$$

Moreover, it is sufficient to verify this inequality for almost all $(a, b) \in J \times J, a \leq b$.

Choose $a \in J$ such that $u(a) \in D(\phi)$. This is possible because the last inclusion is valid for almost all $a \in J$ by (1.13). Without loss of generality we may assume $a = 0$. Now we consider the solution w_ϵ of the Cauchy problem

$$\epsilon w_\epsilon' + w_\epsilon = u, \; w_\epsilon(0) = u(0), \; \epsilon > 0, \tag{1.15}$$

We have the integral representation

$$w_\epsilon(t) = e^{-t/\epsilon} \cdot u(0) + \frac{1}{\epsilon} \int_0^t e^{(s-t)/\epsilon} \cdot u(s)ds. \tag{1.16}$$

It is not difficult to see that $u = \lim w_\epsilon$ in $L^p(0, T_2; V) \cap C([0, T_2]; H)$. The right-hand side of (1.16) is a convex combination of the values of the function $u(s), s \in [0, t]$. So by the convexity of ϕ we have

$$\phi(w_\epsilon(t)) \leq e^{-t/\epsilon} \cdot \phi(u(0)) + \frac{1}{\epsilon} \int_0^t e^{(s-t)/\epsilon} \cdot \phi(u(s))ds.$$

Integrating over $t \in [0, b]$ we obtain the estimate

$$\int_0^b \phi(w_\epsilon(t))dt \leq \epsilon(1 - e^{-b/\epsilon})\phi(u(0)) + \int_0^b (1 - e^{(s-b)/\epsilon}\phi(u(s))ds \leq$$

$$\leq \epsilon\phi(u(0)) + \int_0^b \phi(u(s))ds.$$

Hence $\phi(w_\epsilon) \in L^1(0, b)$ and

$$\limsup_{\epsilon\to 0}\int_0^b \phi(w_\epsilon)dt \leqslant \int_0^b \phi(u)dt.$$

But the lower semicontinuity of Φ (compare with (1.13)) implies

$$\limsup_{\epsilon\to 0}\int_0^b \phi(w_\epsilon)dt \geqslant \int_0^b \phi(u)dt.$$

So

$$\lim_{\epsilon\to 0}\int_0^b \phi(w_\epsilon)dt = \int_0^b \phi(u)dt. \tag{1.17}$$

Now for any test function v (1.2) implies

$$\int_0^b [(v'+Au_n-f_n, v-u_n)+\phi(v)-\phi(u_n)]dt \geqslant \frac{1}{2}|v-u_n|^2\Big|_0^b. \tag{1.18}$$

Since $\lim w_\epsilon = u$ strongly in $L^p(0,b;V)$ and since $\{A(\cdot)u_n(\cdot)\}$ is bounded in $L^{p'}(J;V')$ we have

$$\lim_{\epsilon\to 0}\int_0^b (Au_n, w_\epsilon)dt = \int_0^b (Au, u)dt$$

uniformly in n. So

$$\mathcal{I} = \limsup_{n\to\infty}\int_0^b (Au_n, u_n-u)dt = \limsup_{n\to\infty}\lim_{\epsilon\to 0}\int_0^b (Au_n, u_n-w_\epsilon)dt =$$

$$= \lim_{\epsilon\to 0}\limsup_{n\to\infty}\int_0^b (Au_n, u_n-w_\epsilon)dt.$$

Using (1.18) with $v=w_\epsilon$ we obtain the inequality

$$\mathcal{I} \leqslant \limsup_{\epsilon\to 0}\limsup_{n\to\infty}\left\{\int_0^b (w_\epsilon'-f_n, w_\epsilon-u_n)dt + \right.$$

$$+\int_0^b [\phi(w_\epsilon)-\phi(u_n)]dt-\frac{1}{2}|w_\epsilon-u_n|^2|_0^b\Bigg\},$$

By the convergence properties of $\{u_n\}$ and by (1.13) we have

$$\mathcal{I} \leqslant \limsup_{\epsilon\to 0}\int_0^b (w_\epsilon', w_\epsilon-u)dt-\lim_{\epsilon\to 0}\int_0^b (f_n, w_\epsilon-u)dt+$$

$$+\lim_{\epsilon\to 0}\int_0^b [\phi(w_\epsilon)-\phi(u)]dt-\lim_{\epsilon\to 0}\frac{1}{2}|w_\epsilon-u|^2|_0^b.$$

Using the convergence properties of $\{w_\epsilon\}$ and (1.17) we obtain

$$\mathcal{I} \leqslant \limsup_{\epsilon\to 0}\int_0^b (w_\epsilon', w_\epsilon-u)dt.$$

Now by (1.15) we have

$$\mathcal{I} \leqslant -\limsup_{\epsilon\to 0}\epsilon\int_0^b (w_\epsilon', w_\epsilon')dt \leqslant 0,$$

and so

$$\limsup_{n\to\infty}\int_0^b (Au_n, u_n-u)dt \leqslant 0 \tag{1.19}$$

Since $\mathbf{A}:L^p(0, b; V)\to L^{p'}(0, b; V')$ is a bounded monotone semicontinuous operator (Proposition 2.2 of Ch. 2), it is a pseudo-monotone operator ([44, Proposition 2.5 of Ch. 2]. Then (1.19) implies (1.14). □

It is useful to generalize slightly the previous proposition. Let $A_n(t):V\to V'$, $t\in J$, be monotone semicontinuous operators satisfying condition (c) and inequality (1.3).

COROLLARY 1.7. *Assume that* $\lim A_n(t)=A(t)$ *uniformly with respect to t, in the metric* $d_{V,p}$ *(see (2.8) of Ch. 2), i.e.*

$$\lim_{n\to\infty}\sup_{t\in J, v\in V}\frac{\|A(t)v-A_n(t)v\|_*}{1+\|v\|^{p-1}}=0, \tag{1.20}$$

and assume that $\lim f_n = f$ *in* $L^{p'}(J; V')$. *Let* $u_n \in L^p(J; V) \cap C(J; H)$ *be a solution of* VI (A_n, f_n) *and let* $\lim u_n = u$ *in* $C(J; H)$ *and weakly in* $L^p(J; V)$. *Then* u *is a solution of* VI $(A; f)$.

Proof. It is not difficult to see that u_n is a solution of VI (A, g_n), where $g_n(\cdot) = f_n(\cdot) + A(\cdot)u_n(\cdot) - A_n(\cdot)u_n(\cdot)$. Since $\{u_n\}$ is a bounded subset in $L^p(J; V)$, (1.20) implies that $f = \lim g_n$ in $L^{p'}(J; V')$. Applying Proposition 1.6 we finish the proof. □

The difference between our Proposition 1.6 and standard statements of such kind is the requirement of strong convergence of u_n in $C(J; H)$. We need this in order to pass to the limit in the terms in inequality (1.2) not under the integral sign. In the case of the Cauchy problem such difficulties do not appear. Indeed, the Cauchy problem for inequality (1.2) is equivalent to each of the following problems [118]:

$$\int_0^T [(v' + Au - f, v - u) + \phi(v) - \phi(u)]dt \geqslant \frac{1}{2} | u_0 - v(0) |^2$$

for any test function v, or

$$\int_0^T [(v' + Au - f, v - u) + \phi(v) - \phi(u)]dt \geqslant 0$$

for any test function v such that $v(0) = u_0$ ($u_0 \in \overline{D(\phi)}^H$ is given). In these situations we do not need to control the terms not under the integral sign. In some special cases a passage to the limit similar to ours was made in a more complicated way in [44, Ch. 2, n° 8.3] and [109].

2. Bounded solutions.

Here, unless stated otherwise, we assume that all suppositions of § 1 are valid with $J = \mathbf{R}$ (in particular, condition (c) and estimates (1.3), (1.5)).

2.1. SOME PROPERTIES OF SOLUTIONS.

First we will obtain some estimates for bounded solutions. We will need these in the sequel.

PROPOSITION 2.1. *a) If* $f \in BS^{p'}(\mathbf{R}_+; V')$, *then for any solution* $u(t)$ *of inequality* (1.2) *on* $\mathbf{R}_+$ *we have* $u \in BS^p(\mathbf{R}_+; V) \cap C_b(\mathbf{R}_+; H)$ *and*

$$\| u \|_{C_b(\mathbf{R}_+; H)} \leq C_1(| u(0) |, \| f \|_{S^{p'}}), \tag{2.1}$$

$$\| u \|_{S^p} \leq C_2(| u(0) |, \| f \|_{S^{p'}}), \tag{2.2}$$

where $C_i(x, y)$, $x, y \in \mathbf{R}_+$, $i = 1, 2$, are nonnegative functions, increasing in each variable and depending only on the constants appearing in (1.5) *and the constants of the embedding $V \subset H$.*

b)If $u \in C_B(\mathbf{R}; H)$ is a solution of inequality (1.2) *with $f \in \mathrm{BS}^{p'}(\mathbf{R}; V')$, then $u \in \mathrm{BS}^p(\mathbf{R}; V)$ and*

$$\| u \|_{C_b(\mathbf{R}; H)} \leq C_1(\liminf_{s \to -\infty} | u(s) |, \| f \|_{S^{p'}}), \tag{2.3}$$

$$\| u \|_{S^p} \leq C_2(\liminf_{s \to -\infty} | u(s) |, \| f \|_{S^{p'}}). \tag{2.4}$$

Proof. By the reduction we made in n°. 1.2, we may assume that $\phi \geq 0$. By Lemma 1.4,

$$\frac{1}{2} | u |^2 |_{t_1}^{t_2} + \alpha \int_{t_1}^{t_2} \| u \|^p dt \leq \int_{t_1}^{t_2} \| f \|_* \, \| u \| dt. \tag{2.5}$$

Now we may apply Lemma 1.1 of Ch. 2, setting $\lambda^2 = \frac{1}{2} | u |^2$, $\psi = \| u \|$, $\theta = \| f \|_*$, and $q = p$. Then inequality (2.5) turns into inequality (1.2) of Ch. 2. The constant of the embedding $V \subset H$, say γ, occurs in the estimate $\lambda \leq \sqrt{2}\, \gamma \psi$. This gives estimates of the norms $\| u \|_{C_b}$ and $\| u \|_{S^p}$ in terms of $\| u \|_{L^\infty(0,2,H)}$ and $\| f \|_{S^{p'}}$ similar to (2.1), (2.2). However, $\| u \|_{L^\infty(0,2;H)}$ can be estimated by $| u(0) |$ (see Proposition 1.1), and

$$\| f \|_{L^{p'}(0,2;V')} \leq 2 \| f \|_{S^{p'}}.$$

a) is proved. Part b) is an immediate consequence of a). □

Now we will consider solutions of inequality (1.2) on J, where J is a half-line or $J = \mathbf{R}$. We derive some estimates we need to prove the existence of bounded solutions. We assume in addition that for some $q \geq 2$ one of the following two inequalities is valid for all $v, w \in V$:

$$(A(t)v-A(t)w, v-w) \geqslant \alpha|v-w|^q, \; q \leqslant p, \tag{2.6}$$

$$(A(t)v-A(t)w, v-w) \geqslant \alpha|v-w|^q(1+||v||+||w||)^{p-q}, \; q > p, \tag{2.7}$$

uniformly in t.

PROPOSITION 2.2. *Suppose that inequality* (2.6) *is valid and that* u_1, u_2 *are two solutions of inequality* (1.2) *with common right-hand side* f. *Then for any* $t_0 \in J$ *and all* $t \geqslant t_0$ *(respectively,* $t \leqslant t_0$*),* $t \in J$, *we have the following estimates:*
in the case $q>2$,

$$|u_1(t)-u_2(t)| \underset{(\geqslant)}{\leqslant} \left[\frac{\beta_1}{\beta_2 \cdot |u_1(t_0)-u_2(t_0)|^{q-2} \cdot (t-t_0)+\beta_1}\right]^{1/(q-2)} \cdot \tag{2.8}$$

$$\cdot |u_1(t_0)-u_2(t_0)|;$$

in the case $q=2$,

$$|u_1(t)-u_2(t)| \underset{(\geqslant)}{\leqslant} e^{-\beta_3(t-t_0)} \cdot |u_1(t_0)-u_2(t_0)|. \tag{2.9}$$

Here the constants β_i, $i=1,2,3$, *depend only on* α *and* q.
Proof. Set

$$\lambda^2 = \frac{1}{2}|u_1-u_2|^2.$$

By Lemma 1.5 and estimate (2.6) we have the inequality

$$\lambda^2|_{t_1}^{t_2} + \alpha_1 \int_{t_1}^{t_2} \lambda^q dt \leqslant 0, \; t_1 \leqslant t_2,$$

which coincides with inequality (1.9) of Ch. 2. Applying Lemma 1.3 of Ch. 2 we finish the proof. □

In the case of (2.7) the situation is more complicated and we obtain less precise estimates. But they are sufficient for our purposes.

PROPOSITION 2.3. *Suppose that inequality* (2.7) *is valid and that* u_1, u_2 *are two solutions of inequality (1.2) with common right- hand side f. Set*

$$R = \max\{ \| u_i \|_{C_b(J;H)} \}.$$

Then for any fixed $\delta > 0$ and for any $t_0 \in J$ we have

$$| u_1(t) - u_2(t) | \leq \beta_4 \frac{| u_1(t_0) - u_2(t_0) |^{2/q}}{(t - t_0)^{1/q}}, \ t \geq t_0 + \delta, \tag{2.10}$$

$$| u_1(t) - u_2(t) | \geq \beta_5 (t_0 - t)^{1/2} \cdot | u_1(t_0) - u_2(t_0) |^{q/2}, \ t \leq t_0 - \delta, \tag{2.11}$$

where the constants β_4 and β_5 depend only on α, q, R, and δ.

Proof. Set λ as in the proof of the previous proposition and

$$s(t) = 1 + \| u_1(t) \| + \| u_2(t) \|.$$

By Proposition 2.1,

$$\| s \|_{S^p} \leq C(R).$$

Inequalities (1.10) and (2.7) imply

$$\lambda^2 |_{t_1}^{t_2} + \alpha_1 \int_{t_1}^{t_2} \lambda^q \cdot s^{p-q} dt \leq 0, \ t_1 \leq t_2.$$

Applying Lemma 1.4 of Ch. 2 we finish the proof. □

The following uniqueness theorem is a generalization and improvement of corresponding results in [57, 107].

THEOREM 2.4. *Suppose that one of the inequalities* (2.6) *or* (2.7) *is valid with $q \geq 2$. Then for any $f \in BS^{p'}(\mathbf{R}; V')$ inequality* (1.2) *has at most one solution in the space $C_b(\mathbf{R}; H)$. Moreover, in the case when the estimate* (2.6) *is valid with $q > 2$ there is at most one solution defined on all of* **R**.

Proof. Assume there are two solutions, u_1 and u_2, such that $u_1(t_0) \neq u_2(t_0)$. Then by (2.9) or by (2.11) (the case of (2.6) with $q = 2$ and the case of (2.7)) we conclude immediately that

$$\lim_{t \to -\infty} | u_1(t) - u_2(t) | = +\infty.$$

But this contradicts the boundedness of u_1 and u_2.

In the case of inequality (2.6) with $q>2$, inequality (2.8) implies

$$\lim_{t \to \bar{t}+0} |u_1(t)-u_2(t)| = \infty,$$

where $\bar{t}$ is the root of the denominator in the right-hand side of (2.8). □

The forward estimates we have obtained show that all solutions to our problem come together (in the norm of H) as $t \to +\infty$. The situation is similar to the one arising in the case of regular linear differential operators [24]. But in the last case we have exponential nearness (or, more generally, exponential dichotomy) of solutions, while for our problem we may have polynomial decay of $|u_1-u_2|$. In terms of evolution operators $U(t, t_0)$ (shift along solutions) inequality (2.9) (respectively, (2.8)) means that $U(t, t_0)$ is a (generalized) contraction in the space H for $t>t_0$. Recall that an operator U is said to be a generalized contraction (see, for example, [35]) if

$$|U_v - U_w| \leqslant \rho(\alpha, \beta)|v-w|$$

for all $v, w \in H$ such that $\alpha \leqslant |v-w| \leqslant \beta$, where $\rho(\alpha, \beta)<1$ for $0<\alpha\leqslant\beta<\infty$. Estimate (2.10) shows that $U(t, t_0)$ is only locally Hölder continuous with exponent $2/q$.

Now we will discuss the case when (2.6) is valid only for $v, w \in V$ with $|v|, |w| \geqslant$const. Then, as is not hard to see, the solutions of (1.2) lying in the space $C_b(\mathbf{R}; H)$ form a bounded set. By Proposition 2.1 this set is also bounded in the space $BS^p(\mathbf{R}; V)$. Proposition 1.6 implies that this set is closed in $C_b(\mathbf{R}; H) \cap BS^p(\mathbf{R}; V)$. But in general the question of the existence of such solutions is still open.

2.2. EXISTENCE OF BOUNDED SOLUTIONS.

The following statement is the main result of the present section and gives a partial solution to a problem stated by J.-L. Lions [44, Ch. 4, Problem 10.17].

THEOREM 2.5. *Suppose that condition (c), inequalities* (1.3), (1.5), *and one of the inequalities* (2.6) *or* (2.7) *with* $q \geqslant 2$ *are valid. Then for any* $f \in BS^{p'}(\mathbf{R}; V')$ *there exist a solution* $u \in BS^p(\mathbf{R}; V) \cap C_b(\mathbf{R}; H)$ *of inequality* (1.2), *and this solution is unique.*

Proof. Uniqueness follows from Theorem 2.4.

To prove existence we consider approximate solutions defined by the following Cauchy problem: find a solution

$$u_n \in L^p_{loc}([-n, +\infty); V) \bigcap C([-n, +\infty); H)$$

of inequality (1.2) such that $u_n(-n)=0$. By Proposition 1.1 such solutions are uniquely defined. Set

$$f_n(t) = \begin{cases} f(t), & t \geqslant -n, \\ 0, & u < -n, \end{cases}$$

and extend u_n by zero to the whole axis. Then $u_n(t)$ is a solution of (1.2) with right-hand side f_n.

By Proposition 2.1 we have

$$|| u_n ||_{C_b(\mathbf{R}; H)} \leqslant C, \tag{2.12}$$

$$|| u_n ||_{S^p} \leqslant C, \tag{2.13}$$

where $C>0$ does not depend on n.

Now we will show that $\lim u_n = u$ exists (in some sense). As $f_n(t)=f_m(t)=f(t)$ for $t \geqslant t_{n,m} = -\min\{m, n\}$, (2.12) and Proposition 2.2 and 2.3 (with $t_0 = t_{n,m}$) imply the following estimate:

$$| u_n(t) - u_m(t) | \leqslant \xi_{n,m}(t), \; t \geqslant t_{n,m} \tag{2.14}$$

($t \geqslant t_{n,m} + \delta$ in case (2.7) is valid), where

$$\xi_{n,m}(t) \leqslant \begin{cases} 2Ce^{\beta_3(t_{n,m}-t)}, & q = 2 \leqslant p, \\ 2C \cdot \left[\dfrac{\beta_1}{\beta_2 (2C)^{q-2} \cdot (t - t_{n,m}) + \beta_1} \right]^{1/q-2}, & 2 < q \leqslant p, \\ \beta_4 \dfrac{(2C)^{2/q}}{(t - t_{n,m})^{1/q}}, & q > p. \end{cases} \tag{2.15}$$

As $t_{n,m} \to -\infty$, we conclude from (2.14), (2.15) that $\{u_n\}$ is a Cauchy sequence in the space $C_b([T,\infty);H)$ and, as consequence, in $C(\mathbf{R};H)$. Thus, the required limit exists in the last space.

By (2.13) $\{u_n\}$ is weakly precompact in $L^p_{loc}(\mathbf{R};V)$. So $u=\lim u_n$ weakly in $L^p_{loc}(\mathbf{R};V)$. Passing to the limit in estimates (2.12), (2.13) we see that $u\in BS^p(\mathbf{R};V)\cap C_b(\mathbf{R};H)$, and

$$\|u\|_{C_b(\mathbf{R};H)} \leqslant \liminf \|u_n\|_{C_b(\mathbf{R};H)}, \tag{2.16}$$

$$\|u\|_{S^p} \leqslant \liminf \|u_n\|_{S^p}. \tag{2.17}$$

By Proposition 1.6, u is a solution of (1.2) and the proof is complete. □

REMARK 2.6. Bounds (2.14), (2.15) give rise to the following local estimate for the rate of convergence of the above-constructed approximate solutions:

$$\|u_n-u\|_{C_b([T,\infty);H)} \leqslant \begin{cases} c_1(T)\cdot e^{-\beta n}, & q=2\leqslant p, \\ c_2(T)\cdot n^{-1/(q-2)}, & 2<q\leqslant p, \\ c_3(T)\cdot n^{-1/q}, & q>p. \end{cases}$$

Here the constants $c_j(T)$ depend only on the constant α and on an upper bound of the norm of f in the space $BS^{p'}(\mathbf{R};V')$.

REMARK 2.7. Estimates (2.16), (2.17) and Proposition 2.1 imply that for the above-constructed solution, bounds of the form

$$\|u\|_{C_b(\mathbf{R};H)} \leqslant C_1(\|f\|_{S^{p'}}),$$

$$\|u\|_{S^p} \leqslant C_2(\|f\|_{S^{p'}})$$

are valid. Here $C_j(r)$, $r\geqslant 0$, $j=1,2$, are monotone increasing functions depending only on the constants involved in estimate (1.5).

A useful addition to Theorem 2.5 is the following

PROPOSITION 2.8. *Under the assumptions of Theorem 2.5 the unique bounded solution of* (1.2) *is asymptotically stable in the norm of H.*

Proof. This follows from the forward estimates stated in Propositions 2.3 and 2.4. □

Now we return to the setting of Remark 1.3. Thus, we are given reflexive Banach space V_i, $i=1, \ldots, n$, continuously and densely embedded into a Hilbert space H, operators $A_i(t)$: $V_i \to V_i'$ and a proper convex lower semicontinuous function ϕ on $V = \bigcap V_i$.

By setting

$$A(t)v = \sum_{i=1}^{n} A_i(t)v, \; v \in V,$$

we obtain the operator

$$A(t): V \to V' = \sum_{i=1}^{n} V_i'.$$

Assuming all suppositions of Remark 1.3 being valid, we have

THEOREM 2.9. *Suppose that for some j one of the two inequalities*

$$(A_j(t)v - A_j(t)w, v-w) \geqslant \alpha | v-w |^q, \; v, w \in V_j, \; q \leqslant p_j,$$

or

$$(A_j(t)v - A_j(t)w, v-w) \geqslant \alpha | v-w |^q (1 + || v ||_j + || w ||_j)^{p_j - q},$$

$$v, w \in V_j, \; q > p_j,$$

is valid with $q \geqslant 2$. Then for any

$$f \in \sum \mathrm{BS}^{p_j'}(\mathbf{R}; V_j')$$

there exist a unique solution

$$u \in [\bigcap \mathrm{BS}^{p_j}(\mathbf{R}; V_j)] \bigcap C_b(\mathbf{R}; H)$$

of inequality (1.2).

The *proof* is similar to that of Theorem 2.5, but to construct the approximate solutions we need to use Remark 1.3. Also, to derive estimates of the type of (2.12) and (2.13) a considerable extension of Lemma 1.1 Ch. 2 (the case $p=q$) is needed.

□

Note that we do not need the constants c_1 and c_2 to be independent of t. It is sufficient that they can be chosen uniformly on any finite interval of t's. However, below the uniformity of inequality (1.3) will be fully used.

3. Regularity and almost periodicity of bounded solutions.

3.1. INVERSE OPERATOR.

Under the assumptions of Theorem 2.5 we can consider the inverse operator of our problem, mapping f to the bounded solution u of inequality (1.2). More precisely, given $g\in BS^{p'}(\mathbf{R};V')$ we consider the operator $F_g: h\mapsto u$, where u is the bounded solution of (1.2) with right-hand side $f=g+h$ (u exists by Theorem 2.5). Also, we set $F=F_0$. To be more flexible we introduce a reflexive Banach space E which is intermediate between V and H, i.e. $V\subset E\subset H\subset E'\subset V'$. Moreover, we modify conditions (2.6) and (2.7) in the following way:

$$(A(t)v-A(t)w, v-w) \geqslant \alpha\,||\,v-w\,||_E^q,\ v, w\in V,\ q\leqslant p, \tag{3.1}$$

$$(A(t)v-A(t)w, v-w) \geqslant \alpha\,||\,v-w\,||_E^q(1+||\,v\,||+||\,w\,||)^{p-q},$$

$$v, w\in V,\ q>p, \tag{3.2}$$

where $\alpha>0$ is independent of $t\in\mathbf{R}$. In general these conditions are stronger then (2.6), (2.7), respectively, and coincide with them if $E=H$.

THEOREM 3.1. *Under the assumptions of Theorem 2.5 the operator* $F:BS^{p'}(\mathbf{R};V')\to C(\mathbf{R};H)$ *is continuous. If in addition one of the inequalities* (3.1) *or* (3.2) *is valid with* $q\geqslant 2$, *then the operators*

$$F_g: BS^{\rho'}(\mathbf{R};E')\to C_b(\mathbf{R};H),$$

$$F_g: BS^{\rho'}(\mathbf{R};E')\to BS^{\rho'}(\mathbf{R};E),$$

where $\rho=\min(p,q)$ *and* $g\in BS^{p'}(\mathbf{R};V')$, *are locally Hölderian with exponents* $1/(q-1)$ *and* $2/q(q-1)$, *respectively.*

Proof. We make use of the approximate solutions constructed in n°. 2.2.

1. Assume $h=\lim h_j$ in $BS^{p'}(\mathbf{R}; V')$ and consider approximate solutions u_{nj} and u_n corresponding to the right-hand sides h_j and h, respectively. Applying estimate (1.11) with $t_1=-n, t_2=t, u_1=u_{nj}, u_2=u_n, f_1=h_j$, and $f_2=h$, we obtain the inequality

$$\frac{1}{2}|u_{nj}(t)-u_n(t)|^2 \leqslant$$

$$\leqslant \left[\int_{-n}^{t} \| h_j(\tau)-h(\tau) \|_*^{p'} d\tau\right]^{1/p'} \left[\int_{-n}^{t} \| u_{nj}(\tau)-u_n(\tau) \|^p d\tau\right]^{1/p}.$$

Since $\{u_{nj}\}$ is bounded in $BS^p(\mathbf{R}; V)$, this implies $\lim_j u_{nj}=u_n$ in $C(\mathbf{R}; H)$ for all n. Further, as was stated in n°. 2.2, $\lim_n u_n=u$ and $\lim_n u_{nj}=u_j$ in the space $C(\mathbf{R}; H)$ and this convergence is uniform with respect to j (see Remark 2.6). Changing the order of limits gives $\lim_j u_j=u$ in $C(\mathbf{R}; H)$.

2. **The case of inequality (3.2):** $q>p, \rho=p$. Assume that $h_j \in BS^{p'}(\mathbf{R}; E')$ and $u_i \in BS^p(\mathbf{R}; V) \cap C_b(\mathbf{R}; H)$ are solutions of (1.2) with $f_i=g+h_i, i=1,2$. For corresponding approximate solutions u_{ni} the estimates (1.10) and (3.2) imply

$$\frac{1}{2}|u_{n1}-u_{n2}|^2\Big|_{t_1}^{t_2}+\alpha\int_{t_1}^{t_2}\| u_{n1}-u_{n2} \|_E^q(1+\| u_{n1} \|+\| u_{n2} \|)^{p-q}d\tau \leqslant$$

$$\leqslant \int_{t_1}^{t_2}(h_{n1}-h_{n2}, u_{n1}-u_{n2})d\tau. \tag{3.3}$$

Here $f_{ni}=\chi_n f_i=\chi_n g+\chi_n h_i=g_n+h_{ni}$, where χ_n is the characteristic function of the half-axis $[-n, +\infty)$. By Hölder's inequality with exponents p/q and $p/(p-q)$ (see [2] or (1.2) of Ch. 2) the integral term I on the left-hand side of (3.3) may be estimated as follows:

$$I \geqslant \left[\int_{t_1}^{t_2}\| u_{n1}-u_{n2} \|_E^p d\tau\right]^{q/p} \cdot \left[\int_{t_1}^{t_2}(1+\| u_{n1} \|+\| u_{n2} \|)^p d\tau\right]^{(p-q)/p}.$$

If $0 \leqslant t_2-t_1 \leqslant 3$, then this estimate, the boundedness of the set $\{u_{ni}\}$ in $BS^p(\mathbf{R}; V)$

and the assumption $p<q$ imply the inequality

$$I \geqslant \alpha_1 \left[\int_{t_1}^{t_2} \| u_{n1} - u_{n2} \|_E^p d\tau \right]^{q/p} .$$

So we obtain the estimate

$$\frac{1}{2} | u_{n1} - u_{n2} |^2 \Big|_{t_1}^{t_2} + \alpha_1 \left[\int_{t_1}^{t_2} \| u_{n1} - u_{n2} \|_E^p d\tau \right]^{q/p} \leqslant$$

$$\leqslant \int_{t_1}^{t_2} \| u_{n1} - u_{n2} \|_E \cdot \| h_{n1} - h_{n2} \|_{E'} d\tau \tag{3.4}$$

for $0 \leqslant t_2 - t_1 \leqslant 3$. Since $u_{n1} - u_{n2} \equiv 0$ in a neighbourhood of $-\infty$, we can apply Lemma 1.1 of Ch. 2 with

$$\lambda^2 = \frac{1}{2} | u_{n1} - u_{n2} |^2, \ \psi = \| u_{n1} - u_{n2} \|_E, \ \theta = \| h_{n1} - h_{n2} \|_{E'} .$$

Therefore we obtain

$$| u_{n1}(t) - u_{n2}(t) |^2 \leqslant \tag{3.5}$$

$$\leqslant C \left[\| h_{n1} - h_{n2} \|_{S^{p'}}^{q/(q-1)} + \| h_{n1} - h_{n2} \|_{S^{p'}}^{2/(q-1)} \right],$$

$$\| u_{n1} - u_{n2} \|_{S^p} \leqslant \tag{3.6}$$

$$\leqslant C \cdot \left[\| h_{n1} - h_{n2} \|_{S^{p'}}^{1/q(q-1)} + \| h_{n1} - h_{n2} \|_{S^{p'}}^{2/q(q-1)} \right].$$

Here S^p- and $S^{p'}$-norms are used for functions with values in E and E', respectively. Now we use the evident inequality

$$||h_{n1}-h_{n2}||_{S^{p'}} \leqslant ||h_1-h_2||_{S^{p'}}$$

and the inequalities

$$||u_1-u_2||_{C_b} \leqslant \liminf ||u_{n1}-u_{n2}||_{C_b},$$

$$||u_1-u_2||_{S^p} \leqslant \liminf ||u_{n1}-u_{n2}||_{S^p},$$

which follows from the convergence properties of $\{u_{ni}\}$. By taking into account the fact that $q \geqslant 2$, we obtain from (3.5) and (3.6) the inequalities

$$||u_1-u_2||_{C_b} \leqslant C_1 \cdot ||h_1-h_2||_{S^{p'}}^{1/(q-1)},$$

$$||u_1-u_2||_{S^p} \leqslant C_1 \cdot ||h_1-h_2||_{S^{p'}}^{2/q(q-1)}.$$

These lead to the required statement. We note that in general the constant C_1 depends on upper bounds of the norms of h_i in the space $BS^{p'}(\mathbf{R}; E')$.

3. **The case of inequality (3.1):** $q \leqslant p$, $\rho = q$. This case is similar to the previous one and even simpler (we need not use Hölder's inequality). □

Theorem 3.1 leads to statements on the continuous dependence of solutions on parameters which will be used later on to investigate the regularity and almost periodicity.

Assume that in (1.2) the operator A and the function f depend on an additional parameter $x \in X$, where X is a topological space, i.e. $A = A(t, x)$ and $f = (t, x)$. Assume that for any $x \in X$ all the conditions of Theorem 2.5 are fulfilled with uniform constants in estimates (1.3) and (1.5). Denote by $u(x) = u(t, x)$ the solution of (1.2), which exists according to the theorem. In addition, assume that

$$\lim_{x \to x_0} d_{E,\rho}(x, x_0) = 0, \tag{3.7}$$

where

$$d_{E,\rho}(x, x_0) = \sup_{t \in \mathbf{R}, v \in V} \frac{||A(t,x)v - A(t,x_0)v||_{E'}}{1 + ||v||_E^{\rho-1}}. \tag{3.8}$$

COROLLARY 3.2. *Suppose that for any $x \in X$ one of the inequalities (3.1) or (3.2)* *with* $q\geqslant 2$ *and* α *independent of* x *is valid. Assume* $f(x)=g+h(x)$, *where* $g\in BS^{p'}(\mathbf{R}; V')$, $h(x)\in BS^{p'}(\mathbf{R}; E')$ *and* $\lim_{x\to x_0} h(x)=h(x_0)=0$ *in that space. Then* $\lim_{x\to x_0} u(x)=u(x_0)$ *in the space* $C_b(\mathbf{R}; H)\bigcap BS^p(\mathbf{R}; E)$, *and*

$$||\, u(x)-u(x_0)\, ||_{C_b} \leqslant \tag{3.9}$$

$$\leqslant C\cdot\left[||\, f(x)-f(x_0)\, ||_{S^{p'}}^{1/(q-1)}+d_{E,p}(x, x_0)^{1/(q-1)}\right],$$

$$||\, u(x)-u(x_0)\, ||_{S^p} \leqslant \tag{3.10}$$

$$\leqslant C\cdot\left[||\, f(x)-f(x_0)\, ||_{S^{p'}}^{2/q(q-1)}+d_{E,p}(x, x_0)^{2/q(q-1)}\right],$$

where the constant $c>0$ *depends only on the radius of the ball in* $BS^{p'}(\mathbf{R}; V')$ *which contains* $\{f(x)\}$.

Proof. Here the operator F_g depends on $x\in X$, i.e. $F_g=F_g(x)$. It is easy to see that

$$u(x) = F_g(x_0)[h(x)+A(x)u(x)-A(x_0)] = F_g(x_0)(\hat{h}(x)). \tag{3.11}$$

If x is close to x_0, then $\{u(x)\}$ is bounded in $BS^p(\mathbf{R}; V)$ and, as a consequence, in $BS^p(\mathbf{R}; E)$. This implies $\lim_{x\to x_0} \hat{h}(x)=0$ in $BS^{p'}(\mathbf{R}; E')$. Using the second part of Theorem 3.1 we complete the proof. □

In the case $E=H$ the condition of Corollary 3.2 concerning continuous dependence of f on x in the space $BS^{p'}(\mathbf{R}; H)$ is very restrictive. It turns out that it can be relaxed considerably but in such a way that some continuous dependence of the solution on a parameter is preserved. Namely, representation (3.11) together with the first part of Theorem 3.1 give rise to the following statement.

COROLLARY 3.3. *Suppose that uniformly in* $x\in X$ *one of the two estimates* (2.6) *and* (2.7) *with* $q\geqslant 2$ *is valid. Suppose also that* $\lim_{x\to x_0} f(x)=f(x_0)$ *in* $BS^{p'}(\mathbf{R}; V')$ *and* $\lim_{x\to x_0} d_{V,p}(x, x_0)=0$. *Then* $\lim_{x\to x_0} u(x)=u(x_0)$ *in the space* $C(\mathbf{R}; H)$.

REMARK 3.4. Under the conditions of Corollaries 3.2 or 3.3 we have $\lim_{x\to x_0} u(x)=u(x_0)$ weakly in $L^p_{\mathrm{loc}}(\mathbf{R}; V)$.

3.2. REGULARITY OF SOLUTIONS.

To investigate the problem of time regularity of bounded solutions we need some constructions from the theory of interpolation spaces. A detailed presentation of the topic can be found in [3, 38, 83]. For a Banach couple $E_1 \subset E_0$ (the inclusion is continuous and dense) we denote by $(E_1, E_0)_{\theta,r}$, $0<\theta<1$, $1\leqslant r\leqslant\infty$ the Peetre interpolation functor (in the case $r=\infty$ the values $\theta=0$ and $\theta=1$ are also admissible). We will describe these spaces explicitly in a situation we will use later on. Let $G(s)$ be a strongly continuous semigroup of operators in E_0 and let Λ be its generator. The domain $E_1=D(\Lambda)$ is equipped with the graph norm. The norm in $(E_1, E_0)_{\theta,r}$ is defined by the formulas

$$||v||_{\theta,r} = \begin{cases} \left[\int\limits_0^\infty s^{-r\theta}||(I-G(s))v||_{E_0}^{e}\frac{ds}{s}\right]^{1/r} + ||v||_{E_0}, & 1\leqslant r<\infty, \\ \sup\limits_{s\in(0,\infty)}(s^{-\theta}||(I-G(s))v||_{E_0}) + ||v||_{E_0}, & r=\infty. \end{cases} \tag{3.12}$$

For $C(s)$ we take the group of right shifts

$$[G(s)\psi](\cdot) = \psi(\cdot - s), \quad s\in\mathbf{R},$$

acting in the space $BS^p(\mathbf{R};E)$ for $1\leqslant p<\infty$, or in the space $G_b(\mathbf{R};E)$ for $p=\infty$. It is a group of isometries, but it is not strongly continuous. We denote by UC^p, $1\leqslant p<\infty$, the subspace in $BS^p(\mathbf{R};E)$ (UC_b in $C_b(\mathbf{R};E)$) consisting of the functions for which $G(s)f$ is continuous. It is not difficult to see that UC_b is the space of bounded uniformly continuous functions. Similarly, the space UC^p, $1\leqslant p<\infty$, consists of those $f:\mathbf{R}\to E$ for which the Bochner transform $f^b(t)=f(t+\cdot)$ is a bounded uniformly continuous function with values in $L^p(0,1;E)$. It is clear that UC^p (respectively, UC_b) is closed in $BS^p(\mathbf{R};E)$ (respectively, in $C_b(\mathbf{R};E)$). Moreover, $S^p(\mathbf{R};E)\subset UC^p$, $1\leqslant p<\infty$, and $CAP(\mathbf{R};E)\subset UC_b$. The generator $-\Lambda=-\Lambda_p$ of the group $G(s)$ in UC^p for $1\leqslant p<\infty$, or in UC_b for $p=\infty$, is the operator of differentiation, with dense domain in the corresponding space. We set

$$BS_r^{p,\theta}(\mathbf{R};E) = (D(\Lambda_p), UC^p)_{\theta,r}, \quad 1\leqslant p<\infty,$$

$$C_{b,r}^{\theta}(\mathbf{R};E) = (D(\Lambda_\infty), UC_b)_{\theta,r}.$$

The corresponding norms are denoted by $||\cdot||_{p,\theta,r,E}$ and $||\cdot||_{\theta,r,E}$, respectively (the symbol E is omitted if the space E is clear from the context).

Now we give a more direct description of the spaces introduced (here and below the case $r=\infty$, $\theta=0$ is excluded). The space $BS_r^{p,\theta}(\mathbf{R};E)$ (respectively, $C_{b,r}^{\theta}(\mathbf{R};E)$) consists of those functions v for which the norm (3.12) with $E_0=BS^p(\mathbf{R};E)$ (respectively, with $E_0=C_b(\mathbf{R};E)$) is finite. Indeed, we set

$$\Phi_{\theta,r}(\phi) = \left[\int_0^{\infty}(t^{-\theta}\phi(t))^r\frac{dt}{t}\right]^{1/r}, \ \phi \geq 0$$

(with an evident change in case $r=\infty$). A direct calculation shows that for $\phi_s(t)=\phi(t/s)$ we have

$$\Phi_{\theta,r}(\phi_s) = s^{-\theta}\Phi_{\theta,r}(\phi).$$

So, if $\Phi_{\theta,r}(\phi)<\infty$, then $\phi_s\to 0$ as $s\to\infty$ in the space $L^r(\mathbf{R}_+, t^{-\theta r+1}dt)$ and, as a consequence, almost everywhere. Thus we conclude that $\lim_{t\to 0}\phi(t)=0$. Applying this to $\phi(t)=||(I-G(t))v||_{E_0}$ we obtain that finiteness of the norm (3.12) implies

$$\lim_{t\to 0}||(I-G(t))v||_{E_0} = 0.$$

Since the $G(s)$ are isometries, we have

$$\lim_{t\to 0}||(G(s)-G(s+t))v||_{E_0} = \lim_{t\to 0}||(I-G(t))v||_{E_0} = 0,$$

i.e. $v\in UC^p$ (respectively, $v\in UC_b$).

Membership of a function to the spaces we have introduced may be interpreted as boundedness of it and of some "fractional derivative".

THEOREM 3.5. *Suppose that the conditions of Theorem* 3.1 *are satisfied and that for some* γ, $0<\gamma\leq 1$, *we have*

$$||A(\cdot)v-A(\cdot+s)v||_{E'} \leq C\cdot\min(1,s^{\gamma})\cdot(1+||v||_E^{\rho-1}), \tag{3.13}$$

$$s\in\mathbf{R}_+,\ v\in V,$$

where $\rho=\min(p,q)$. *Let* $f\in BS_r^{p',\theta}(\mathbf{R};E')$, *where* $1\leq r\leq\infty$, $0<\theta<\gamma$ *for* $r<\infty$, *and* $0<\theta\leq\gamma$ *for* $r=\infty$. *Then the unique bounded solution of inequality (1.2) belongs to the space*

$$\mathrm{BS}_l^{p,\eta}(\mathbf{R};E)\cap C^{\mu}_{b\nu}(\mathbf{R};H),$$

where $l=rq(q-1)/2$, $\eta=2\theta/q(q-1)$, $\nu=r(q-1)$, *and* $\mu=\theta/(q-1)$.

Proof. We introduce the notation

$$[\Delta(s)\psi](\cdot) = \psi(\cdot)-\psi(\cdot-s) = [(I-G(s))\psi](\cdot)$$

and use Corollary 3.2, setting $X=\mathbf{R}_+, A(t,s)=A(t-s)$ and $f(t,s)=f(t-s)$ for $s\in\mathbf{R}_+$. Since the bounded solution is unique, we have $u(t,s)=u(t-s)$. Inequality (3.13) implies

$$d_{E,p}(s,0) \leqslant C\cdot\min(1,s^{\gamma}).$$

Now, by (3.9) and (3.10), we obtain the estimates

$$|\,[\Delta(s)u](\cdot)\,| \leqslant C\cdot[\,\|\Delta(s)\|_{S^{p'}}^{1/(q-1)}+\min(1,s^{\gamma/(q-1)})], \tag{3.14}$$

$$\|\Delta(s)u\|_{S^{p}} \leqslant C\cdot[\,\|\Delta(s)\|_{S^{p'}}^{1/q(q-1)}+\min(1,s^{2\gamma/q(q-1)})]. \tag{3.15}$$

Let $r<\infty$. We raise inequality (3.14) to the power ν and multiply it by $s^{-\nu\mu-1}$. We treat (3.15) similarly, replacing ν by l and $s^{-\nu\mu-1}$ by $s^{-\eta l-1}$. Integrating the result and taking into account the relations which define l,η,μ, and ν, we obtain the estimates

$$\|u\|^{\nu}_{\mu,\nu} \leqslant C\cdot[\,\|f\|^{r}_{p',\theta,r}+\int_0^1 s^{r\gamma-r\theta-1}ds+\int_1^{\infty} s^{-r\theta-1}ds],$$

$$\|u\|^{l}_{p,\eta,l} \leqslant C\cdot[\,\|f\|^{r}_{p',\theta,r}+\int_0^1 s^{r\gamma-r\theta-1}ds+\int_1^{\infty} s^{-r\theta-1}ds].$$

The first integral on the right-hand side convergent for $\theta<\gamma$ (the second one always converges, since $r\theta>0$), and the required statement is proved for $r<\infty$.

The case $r=\infty$ is treated similarly, by taking the supremum instead of integrating. □

Now we give a more explicit description of the spaces $BS^{p;1}_{\infty}$ and $C^1_{b;\infty}$. First of all we present some properties of averaging operators. Let $\psi \in C_0^\infty(\mathbf{R})$, $\psi \geqslant 0$, $\operatorname{supp}\psi \subset (-1, 1)$, and

$$\int_{\mathbf{R}} \psi(t)dt = 1.$$

We set $\psi_\delta(t) = \delta^{-1}\psi(t/\delta)$, $\delta > 0$. For any function $v \in L^1_{\mathrm{loc}}(\mathbf{R}; E)$ the functions

$$v_\delta = \psi_\delta \star v \tag{3.16}$$

are well-defined.

LEMMA 3.6. *a) For $v \in L^\infty(\mathbf{R}; E)$ we have*

$$\| v_\delta \|_{L^\infty} \leqslant \| v \|_{L^\infty}, \tag{3.17}$$

b) For $v \in BS^p(\mathbf{R}; E)$ we have

$$\| v_\delta \|_{S^p} \leqslant \| v \|_{S^p}. \tag{3.18}$$

c) If $v \in UC_b$, then

$$\lim_{\delta \to 0} \| v_\delta - v \|_{C_b} = 0. \tag{3.19}$$

d) If $v \in UC^p$, then

$$\lim_{\delta \to 0} \| v_\delta - v \|_{S^p} = 0. \tag{3.20}$$

e) If $v \in BS^1(\mathbf{R}; E)$, then $v_\delta \in C_b^\infty(\mathbf{R}; E)$.

Proof. a) and c) are well-known (see, for example, [48]). To prove b) and d) we consider the Bochner transform $BS^p(\mathbf{R}; E) \to C_b(\mathbf{R}; BS^p(\mathbf{R}; E))$, $v \mapsto v^b$. We have

$$(\psi_\delta \star v^b)(t) = \int \psi_\delta(x) v^b(t-x)dx = \int \psi_\delta(x) v(t+\eta-x)dx = (\psi_\delta \star v)^b(t),$$

i.e. the *averaging operator commutes with the Bochner transform*. The last transform is isometric, therefore a) and c) imply b) and d), respectively.

To prove d) we note that

$$v_\delta^{(n)}(t) = \int_{\operatorname{supp}\psi_\delta} \psi_\delta^{(n)}(x) v(t-x)dx.$$

Hence

$$||v_\delta^{(n)}(t)||_E \leqslant ||\psi_\delta^{(n)}||_{C_b} \int_{\mathrm{supp}\,\psi_\delta} ||v(t-x)||_E dx \leqslant ||\psi_\delta^{(n)}||_{C_b} \cdot ||v||_{S^1}.$$

LEMMA 3.7. *We have the following topological identifications:*

$$C^1_{b,\infty}(\mathbf{R};E) = \{v \in C_b(\mathbf{R};E) \mid v' \in L^\infty(\mathbf{R};E)\}, \tag{3.21}$$

$$\mathrm{BS}^{p,1}_\infty(\mathbf{R};E) = \{v \in \mathrm{BS}^p(\mathbf{R};E) \mid v' \in \mathrm{BS}^p(\mathbf{R};E)\}. \tag{3.22}$$

Here the derivative is regarded in the sense of distributions and the right-hand sides are equipped with the graph norms.

Proof. First of all we recall that the relative completion of the space E_1 with respect to the space E_0 is the Banach space $E_{01} = \{v \in E_0 \mid v = \lim v_n$ in E_0, $||v_n|| \leqslant R$ for some $R\}$. The norm is defined by the formula $||v||_{E_{01}} = \inf R$. It is well-known (see, for example, [38]) that the space $(E_1, E_0)_{1,\infty}$ coincides with the relative completion of E_1 with respect to E_0. This and the fact that Λ is a differentiation operator imply that the left-hand sides of (3.21) and (3.22) are contained in the right-hand sides, respectively. Now we prove the opposite inclusion. Let $v \in C_b(\mathbf{R};E)$ and $v' \in L^\infty(\mathbf{R};E)$. It is obvious that $v \in \mathrm{UC}_b$. By Lemma 3.6 e),

$$v_\delta \in C_b^\infty(\mathbf{R};E) \subset D(\Lambda_\infty).$$

Since the operator of differentiation and the operator of averaging commute, by Lemma 3.6 a), c) we have

$$v = \lim v_\delta \text{ in } C_b(\mathbf{R};E),$$

$$||v_\delta'||_{L^\infty} \leqslant ||v'||_{L^\infty}.$$

Hence $v \in C^1_{b,\infty}(\mathbf{R};E)$. The proof of (3.22) is similar by using Lemma 3.6. b), d).

COROLLARY 3.8. *Under the conditions of Theorem 3.5, let $q=2, \gamma=1, f \in \mathrm{BS}^{p'}(\mathbf{R};E)$ and $f' \in \mathrm{BS}^{p'}(\mathbf{R};E')$. Then for the unique solution $u \in \mathrm{BS}^p(\mathbf{R};V) \cap C_b(\mathbf{R};H)$ of (1.2) we have $u' \in \mathrm{BS}^p(\mathbf{R};E) \cap L^\infty(\mathbf{R};H)$, and*

$$(u'(t)+A(t)u(t)-f(t),\, v-u(t))+\phi(v)-\phi(u(t)) \geqslant 0 \tag{3.23}$$

for all $v \in V$ and almost all $t \in \mathbf{R}$, i.e. u is a strong solution of the variational inequality.

Proof. To obtain the first part it is sufficient to apply Theorem 3.5 with $r=\infty$, $\theta=1$, and to use Lemma 3.7. Now the second part is a consequence of standard results concerning strong solutions of variational inequalities (see [44, 118]). □

3.3. ALMOST PERIODICITY.

Our approach to almost periodicity of solutions is based on statements concerning continuous dependence of solutions on parameters and, essentially, is a realization of Favard's principle (especially in the form contained in [93]).

We assume valid all suppositions of § 2, with the following condition instead of (c):

($\mathrm{cap}_{V,p}$) *The family of functions $\{(1+\|v\|^{p-1})^{-1}A(\cdot)v,\, v \in V\}$ with values in V' is equicontinuous and uniformly almost periodic.*

Obviously, condition (c) follows from this. By Proposition 2.4 of Ch. 2 we may assume that the family $A(t)$ of monotone semicontinuous operators is defined for all $t \in \mathbf{R}_B$. Moreover, any inequality (1.3), (1.5), (2.6), (2.7), (3.1), or (3.2) will be valid for all $t \in \mathbf{R}_B$, provided it is valid for all $t \in \mathbf{R}$ with uniform constants.

THEOREM 3.9. *Suppose that condition* ($\mathrm{cap}_{V,p}$) *and estimates (1.3), (1.5) and (2.1) or (2.2) with $q \geqslant 2$ are satisfied. If $f \in S^{p'}(\mathbf{R}; V')$, then the unique bounded solution $u(t)$ of inequality* (1.2) *belongs to* $\mathrm{CAP}(\mathbf{R}; H)$. *If, moreover, $A(t)=A_0+A_1(t)$,* (3.1) *or* (3.2) *holds (with $q \geqslant 2$), $f \in S^{\rho'}(\mathbf{R}; E')$, where $\rho=\min(p, q)$, and $A_1(\cdot)$ satisfies condition* ($\mathrm{cap}_{E,\rho}$), *then $u \in \mathrm{CAP}(\mathbf{R}; H) \cap S^{\rho}(\mathbf{R}; E)$.*

Proof. Along with (1.2) we consider the family of inequalities VI $(A(\cdot - s), f(\cdot + s))$ depending on a parameter $s \in \mathbf{R}_B$:

$$\int_{t_1}^{t_2} [(v' + A(t+s)u_s - f(t+s),\, v(t)-u_s(t)) + \phi(v) - \phi(u_s)]dt \geqslant \tag{3.24}$$

$$\geqslant \frac{1}{2}|v-u_s|^2\Big|_{t_1}^{t_2}$$

for any test function $v(t)$. Every one of these inequalities has a unique solution $u_s \in C_b(\mathbf{R}; H) \cap BS^p(\mathbf{R}; V)$ and, by uniqueness,

$$u_s(\cdot) = u(\cdot+s),\ s\in\mathbf{R}. \tag{3.25}$$

Now we note that $f(\cdot+s)$ is a continuous function of s with values in $BS^{p'}(\mathbf{R}; V')$ (see Ch. 1, n°. 4.1). By condition $(\mathrm{cap}_{V,p})$ we have $\lim_{s\to s_0} d_{V,p}(s, s_0)=0$ for any $s_0\in\mathbf{R}_B$. Hence, by Corollary 3.3 the solution u_s depends continuously on $s\in\mathbf{R}_B$ in the topology of the space $C(\mathbf{R}; H)$. Setting $u(s)=u_s(0)$, $s\in\mathbf{R}_B$, we obtain a well-defined (by (3.25)) continuous extension of $u(t)$ to $\mathbf{R}_B$. Thus $u\in\mathrm{CAP}(\mathbf{R}; H)$ and the first part of the Theorem is proved.

To prove the second part we note that Corollary 1.2 is applicable in this situation. Hence u_s depends continuously on $s\in\mathbf{R}_B$ in the topology of $BS^p(\mathbf{R}; E)$. Therefore, by (3.25), $u(\cdot+s)$ has a continuous extension to $\mathbf{R}_B$ with values in $BS^p(\mathbf{R}; E)$. But this just means that $u\in S^p(\mathbf{R}; E)$. □

REMARK 3.10. Under the assumptions of the first part of Theorem 3.9 the solution $u(t)$ is also a weakly S^p-a.p. function with values in V. This follows from Proposition 5.12 of Ch. 1. □

Now we will show that the assertion of Corollary 3.3 on continuous dependence of solutions on parameters may be strengthened, provided the solutions are almost periodic.

PROPOSITION 3.11. *Assume that the operators* $A(t, x): V\to V'$, $t\in\mathbf{R}$, $x\in X$, *satisfy for every* x *condition* $(\mathrm{cap}_{V,p})$ *and inequalities (1.3), (1.5) and (2.6) or (2.7), where* $q\geqslant 2$ *(with the constants independent of* t *and* x*). Let* $\lim_{x\to x_0} f(x)=f(x_0)$ *in* $S^{p'}(\mathbf{R}; V')$ *and* $\lim_{x\to x_0} d_{V,p}(x, x_0)=0$. *Then* $\lim_{x\to x_0} u(x)=u(x_0)$ *in* $\mathrm{CAP}(\mathbf{R}; H)$.

Proof. We consider inequality (3.24), which in our case depends on parameters $s\in\mathbf{R}_B$ and $x\in X$. Under our conditions Corollary 3.3 (with X replaced by $\mathbf{R}_B\times X$) is applicable. So, if $(s, x)\in\mathbf{R}\times X$ and $(s, x)\to(s_0, x_0)$ in $\mathbf{R}_B\times X$, then

$$\lim u(\cdot+s, x) = u_{s_0}(\cdot, x_0) \tag{3.26}$$

in $C(\mathbf{R}; H)$. Now we show that $\lim u(t, x)=u(t, x_0)$ as $x\to x_0$ in X, uniformly in $t\in\mathbf{R}$. If not, then there are nets $\{s_\gamma\}\subset\mathbf{R}$ and $\{x_\gamma\}\subset X$ such that $\lim x_\gamma=x_0$ and

$$|\, u(s_\gamma, x_\gamma)-u(s_\gamma, x_0)\,| \geqslant \epsilon \tag{3.27}$$

for some $\epsilon>0$. By the compactness of $\mathbf{R}_B$ we may assume that $\lim s_\gamma=s_0\in\mathbf{R}_B$. Since $u(\cdot, x_0)\in\mathrm{CAP}(\mathbf{R}; H)$, inequality (3.27) implies

$$|\, u(s_\gamma, x_\gamma)-u_{s_0}(0, x_0)\,| \geqslant$$

$$\geqslant |u(s_\gamma, x_\gamma) - u(s_\gamma, x_0)| - |u(s_\gamma, x_0) - u_{s_0}(0, x_0)| \geqslant \epsilon/2$$

for large γ. This contradicts (3.26). □

Now we will briefly discuss the regularity of almost periodic solutions. For this purpose we set

$$S_r^{p,\theta}(\mathbf{R}; E) = S^p(\mathbf{R}; E) \bigcap \mathrm{BS}_r^{p,\theta}(\mathbf{R}; E).$$

It is not difficult to see that this space coincides with $(E_1, E_0)_{\theta,r}$, where $E_0 = S^p(\mathbf{R}; E)$ and E_1 is the domain of the generator of the group of right shifts in E. Similarly, we set

$$\mathrm{CAP}_r^\theta(\mathbf{R}; E) = \mathrm{CAP}(\mathbf{R}; E) \bigcap C_{b,r}^\theta(\mathbf{R}; E).$$

For $r < \infty$ the space Trig($\mathbf{R}; E$) is dense both in $S_r^{p,\theta}(\mathbf{R}; E)$ and in $\mathrm{CAP}_r^\theta(\mathbf{R}; E)$ (for $r = \infty$ this is not true). Indeed, in our situation E_1 is dense in $S_r^{p,\theta}(\mathbf{R}; E)$ for general reasons (see, for example, [3, Theorem 3.4.2]). Further, Trig($\mathbf{R}; E$)$\subset E_1$ and the embedding is dense in the graph norm, because the Bochner-Fejér operators commute with the group of shifts and, as consequence, with its generator. It remains to observe that the graph norm in E_1 is stronger than the norm of $S_r^{p,\theta}(\mathbf{R}; E)$. The case of CAP_r^θ is similar.

Theorem 3.5 and 3.9 immediately imply

PROPOSITION 3.12. *Let $A(t) = A_0 + A_1(t)$. Assume that $A(t)$ satisfies inequalities* (1.3), (1.5) *and* (3.1) *or* (3.2) *with $q \geqslant 2$, and that $A_1(t)$ satisfies condition* ($\mathrm{cap}_{E,\rho}$) *and inequality* (3.13), *where $\rho = \min(p, q)$ and $0 < \gamma \leqslant 1$. Let $f \in S_r^{p',\theta}(\mathbf{R}; E')$, where $1 \leqslant r \leqslant \infty$, $0 < \theta < \gamma$ for $r < \infty$, and $0 < \theta \leqslant \gamma$ for $r = \infty$. Then the unique a.p. solution of inequality* (1.2) *lies in the space $S_l^{p,\eta}(\mathbf{R}; E) \bigcap \mathrm{CAP}_\nu^\mu(\mathbf{R}; H)$, where $l = rq(q-1)/2$, $\eta = 2\theta / q(q-1)$, $\nu = r(q-1)$, and $\mu = \theta(q-1)$.*

Now note that in the case $q = 2$, $\gamma = 1$, $f \in S^{p'}(\mathbf{R}; E')$, $f' \in S^{p'}(\mathbf{R}; E')$ Corollary 3.8 gives $u' \in \mathrm{BS}^p(\mathbf{R}; E) \bigcap L^\infty(\mathbf{R}; H)$. It does not imply, though, that $u' \in S^p(\mathbf{R}; E) \bigcap \mathrm{CAP}(\mathbf{R}; H)$. Below, however, we will show that u' is a.p. in the sense of Besicovitch (see n°. 5.3).

REMARK 3.13. All previous results can be extended to the situation described in Remark 1.3 (compare with Theorem 2.9).

REMARK 3.14. Under condition (3.1) Proposition 2.8 may be strengthened in the following way. Let u_1 and u_2 be two solutions of inequality (1.2) defined on $\mathbf{R}_+$.

Then (1.10) implies $u_1 - u_2 \in L^p(\mathbf{R}_+; E)$.

4. The use of compactness.

In this section we make the additional assumption that the embedding $V \subset H$ is compact.

4.1. BOUNDED SOLUTIONS.

We consider a subset

$$U = U(c_1, c_2, \alpha, \beta, r) \subset C_b(\mathbf{R}; H) \bigcap \mathrm{BS}^p(\mathbf{R}; V),$$

defined in the following way. Any function $u \in U$ satisfies some inequality $VI(A, f)$ (depending, in general, on u), such that A satisfies condition (c) and inequalities (1.3), (1.5) with fixed constants c_1, c_2, α, and β, and $\| f \|_{S^{p'}} \leqslant r$ (the convex function ϕ is fixed).

LEMMA 4.1. *Any $C_b(\mathbf{R}; H)$-bounded subset of U is precompact in $C(\mathbf{R}; A)$.*
Proof. We set

$$U_R = \{u \in U\} \mid \| u \|_{C_b} \leqslant \mathbf{R}\}.$$

By Proposition 2.1 , $U_R$is bounded in $\mathrm{BS}^p(\mathbf{R}; V)$, i.e.

$$\| u \|_{S^p} \leqslant R_1, \ u \in U_R. \tag{4.1}$$

We prove that the set U_R is uniformly equicontinuous in the norm of H. Fix $\epsilon > 0$ and suppose that for $u \in U_R$, $t_0 \in \mathbf{R}$ and any $\delta \in [0, \delta_0]$ we have the inequality

$$| u(t_0 + \delta) - u(t_0) |^2 \leqslant \epsilon.$$

Since u is a solution of an inequality of the type (1.2), by (1.10) we have

$$| u(t+\delta) - u(t) |^2 \leqslant | u(t_0+\delta) - u(t_0) |^2 +$$

$$+2\int_{t_0}^{t} (g_1(\tau) - g_2(\tau), u(\tau) - u(\tau+\delta))d\tau, \ t \geqslant t_0, \tag{4.2}$$

where $g_1(t)=f(t)-A(t)u(t)$ and $g_2(t)=g(t+\delta)$. By (1.3) and (4.1) we obtain from (4.2) that for some $\theta>0$, independent of $t_0\in\mathbf{R}$, $\delta_0>0$, and $u\in U_R$, the inequality

$$|u(t+\delta)-u(t)|^2 \leqslant 2\epsilon,\ \delta\in[0,\delta_0],\ t\in[t_0,t_0+\theta],\ u\in U_R, \tag{4.3}$$

is valid.

Let $a_k=k\theta/2$, $k\in\mathbf{Z}$. Then (we may assume that $\theta\leqslant 2$) (4.1) implies

$$\int_{a_{k-1}}^{a_k} \|u(t)\|^p dt \leqslant R_1^p,\ u\in U_R.$$

Further, by Lemma 1.4 $\{\phi(u)\,|\,u\in U_R\}$ is a bounded subset in $\mathrm{BS}^1(\mathbf{R})$ and so

$$\int_{a_{k-1}}^{a_k} \phi(u(t))dt \leqslant R_2,\ u\in U_R.$$

Here R_2 depends only on R and on the constants involved in (1.3) and (1.5). Hence there are $R_3>0$, depending only on R_1 and R_2, and $\tau_k\in[a_{k-1},a_k]$, depending on u, in general, such that

$$\|u(\tau_k)\| \leqslant R_3,\ \phi(u(\tau_k)) \leqslant R_3. \tag{4.4}$$

Now in (1.2) we set $t_1=\tau_k$, $t=\tau_k+\delta$ and $v(t)\equiv u(\tau_k)$. Then we obtain

$$|u(\tau_k+\delta)-u(\tau_k)|^2 \leqslant$$

$$\leqslant 2\int_{\tau_k}^{\tau_k+\delta} [(A(t)u(t)-f(t),\ u(\tau_k)-u(t))+\phi(u(\tau_k))-\phi(u(t))]dt.$$

This and inequalities (1.3), (4.1) and (4.4) imply the existence of $\delta_0>0$, independent on u and k, such that

$$|u(\tau_k+\delta)-u(\tau_k)|^2 \leqslant \epsilon,\ \delta\in[0,\delta_0],\ k\in\mathbf{Z},\ u\in U_R.$$

Now by (4.3) with t_0 replaced by τ_k we have

$$|u(t+\delta)-u(t)|^2 \leqslant 2\epsilon,\ \delta\in[0,\delta_0],\ t\in[\tau_k,\tau_k+\theta],\ u\in U_R.$$

Since $[a_k, a_{k+1}] \subset [\tau_k, \tau_k + \theta]$ and $k \in \mathbf{Z}$ is arbitrary, the required equicontinuity of U_R is proved.

To complete the proof we recall the following statement [19] (see, also [11, Ch. IV, Theorem 4.1]):

if a subset of $C([a, b]; H) \cap L^p(a, b; V)$ is bounded in $L^p(a, b; V)$ and equicontinuous in the norm of H, then it is precompact in $C([a, b]; H)$, provided the embedding $V \subset H$ is compact. □

REMARK 4.2. In the previous argument inequality (1.5) was used only to prove the boundedness of U_R in the space $\mathrm{BS}^p(\mathbf{R}; V)$. So, if we consider the set $U_1(c_1, c_2, R, r)$ consisting of the solutions $u \in C_b(\mathbf{R}; H) \cap \mathrm{BS}^p(\mathbf{R}; V)$ with $\| u \|_{S^p} \leqslant R$, $\| u \|_{C_b} \leqslant R$, of all inequalities $VI(A, f)$ for which $\| f \|_{S^{p'}} \leqslant r$ and for which (1.3) is satisfied, then this set is precompact in $C(\mathbf{R}; H)$ too. There are also similar statements on compactness of certain sets of solutions of Cauchy problems for variational inequalities.

Now we are able to establish the following theorem, concerning existence of bounded solutions.

THEOREM 4.3. *Suppose that the embedding $V \subset H$ is compact and that the operators $A(t): V \to V'$ satisfy condition (c) and inequalities (1.3), (1.5). Then for any $f \in \mathrm{BS}^{p'}(\mathbf{R}; V)$ there is a solution $u \in C_b(\mathbf{R}; H) \cap \mathrm{BS}^p(\mathbf{R}; V)$ of inequality* (1.2).

Proof. Along with inequality (1.2) we consider the following approximate problem: find $u_\lambda \in L^p_{\mathrm{loc}}(\mathbf{R}; V) \cap C(\mathbf{R}; H)$, where $\lambda > 0$, such that

$$\int_{t_1}^{t_2} [(v' + Au_\lambda + \lambda u_\lambda - f, v - u_\lambda) + \phi(v) - \phi(u_\lambda)]dt \geqslant \frac{1}{2} | v - u_\lambda |^2 \Big|_{t_1}^{t_2}, \tag{4.5}$$

$$t_1 \leqslant t_2,$$

for all test functions v (the integral containing λu_λ is meaningful since $v \in C(\mathbf{R}; H)$. For $f \in \mathrm{BS}^{p'}(\mathbf{R}; V)$ this inequality has a unique solution $u_\lambda \in C_b(\mathbf{R}; H) \cap \mathrm{BS}^p(\mathbf{R}; V)$. Indeed, for $p \geqslant 2$ we can apply Theorem 2.5, while in the case $1 < p < 2$ we must use Theorem 2.9 with $V_1 = V$, $V_2 = H$. By Remark 2.7 we conclude that the set $\{u_\lambda\}$ is bounded in $C_b(\mathbf{R}; H) \cap \mathrm{BS}^p(\mathbf{R}; V)$. Therefore, by Lemma 4.1, this set is precompact in the space $C(\mathbf{R}; H)$. So, for some sequence $\lambda_i \to 0 \lim u_{\lambda_i} = u$ exists in $C(\mathbf{R}; H)$. Since the space V is reflexive, we may suppose, in addition, that $u_{\lambda_i} \to u$ weakly in $L^p_{\mathrm{loc}}(\mathbf{R}; V)$. It is also not difficult to see that $u \in C_b(\mathbf{R}; H) \cap \mathrm{BS}^p(\mathbf{R}; V)$.

The passage to the limit in (4.5), using Proposition 1.6 and the obvious equality $\lim \lambda u_\lambda = 0$ in $C(\mathbf{R}; H)$ show that u is a solution of (1.2). □.

REMARK 4.4. To prove Theorem 4.3 we may use the approximations introduced in n°. 2.2 instead of (4.5) (this was done in the paper [57]). But approximations (4.5) seems to be preferable. Moreover, they will appear in § 5, when dealing with Besicovitch a.p. solutions.

PROPOSITION 4.5. *Under the conditions of Theorem* 4.3, *for any solution* $u \in C_b(\mathbf{R}; H) \cap BS^p(\mathbf{R}; V)$ *the set* $u(\mathbf{R})$ *is precompact in H.*

Proof. For any $s \in \mathbf{R}$ the function $u(\cdot+s)$ satisfies the inequalities VI $(A(\cdot+s), f(\cdot+s))$. So, $\{u(\cdot+s)\} \subset U$. Since $||f(\cdot+s)||_{S^{p'}}$ and $||u(\cdot+s)||_{C_b} = ||u||_{C_b}$, Lemma 4.1 applies and we finish the proof. □.

In the case of equation (1.6) ($\phi \equiv 0$), Theorem 4.3 is a result obtained by V. Zhikov [22, 25]. Zhikov used usual compactness arguments based on boundedness of derivatives of solutions in $BS^{p'}(\mathbf{R}; V')$ (which is evident in this case). In our situation such arguments must be replaced by the more delicate Lemma 4.1.

4.2. ALMOST PERIODICITY OF SOLUTIONS.

Suppose that condition ($\text{cap}_{V,p}$) and inequalities (1.3), (1.5) are satisfied. We will obtain an existence theorem for a.p. solutions which is an extension of corresponding result of V. Zhikov [22, 25] concerning equation (1.6). To prove it we will use a suitable version of arguments from [25] (modulo Proposition 1.6 and results of n°. 4.1 it is even somewhat simpler than the original one).

To begin we state

PROPOSITION 4.6. *For any two solutions* $u_1, u_2 \in BS^p(\mathbf{R}; V) \cap C_b(\mathbf{R}; H)$ *of* (1.2) *with common right-hand side* $f \in S^{p'}(\mathbf{R}; V')$ *the invariance identity*

$$|u_1(t) - u_2(t)| \equiv const, \; t \in \mathbf{R}, \tag{4.6}$$

is valid.

Proof. We consider the set

$$S = S(R) \subset C_b(\mathbf{R}; H)$$

of solutions of (1.2) with norms $\leqslant R$. By Proposition 1.6, S is closed in $C(\mathbf{R}; H)$ and, by Lemma 4.1, it is precompact in that space. So S is compact. This implies compactness of the set

$$K = \{y \in H \mid y = u(0),\ u \in S\}$$

in the space H.

Now we find a net $\{s_\gamma\} \subset \mathbf{R}$ such that $s_\gamma \to -\infty$ in $\mathbf{R}$ and $s_\gamma \to 0$ in $\mathbf{R}_B$. (This is possible in the view of the structure of $\mathbf{R}_B$, described in Ch. 1.) Then $\lim f(\cdot + s) = f(\cdot)$ in $S^{p'}(\mathbf{R}; V')$ and, by condition ($\mathrm{cap}_{V,p}$), $\lim A(\cdot + s_\gamma) = A(\cdot)$ in the metric $d_{V,p}$ (see, Ch. 2, § 2), uniformly on $\mathbf{R}$. Now, for any $y \in K$ we choose some member $v \in S$ such that $v(0) = y$, and define maps $P_\gamma: K \to H$ by the rule $P_\gamma y = v(s_\gamma)$. The net of maps $\{P_\gamma\}$ is precompact in the topology of pointwise convergence. Let P be a limit point of $\{P_\gamma\}$. Then for any $y_1, y_2 \in K$,

$$w_i = Py_i = \lim v_i(s_{\gamma'}),\ i = 1, 2,$$

where $\{s_{\gamma'}\} \subset \{s_\gamma\}$ is a subset depending on y_1, y_2, Since $v_i(\cdot + s_\gamma)$ satisfies inequality VI $(A(\cdot + s), f(\cdot + s_\gamma))$, we have $v_i(\cdot + s_\gamma) \in U_R$. Hence we may assume that $\lim v_i(\cdot + s_{\gamma'}) = u_i(\cdot)$ exists in $C(\mathbf{R}; H)$ and exists weakly in $L^p_{\mathrm{Loc}}(\mathbf{R}; V)$, and, moreover, $u_i(0) = w_i$ $(i = 1, 2)$. By Corollary 1.7, u_1 and u_2 are solutions of VI (A, f). Further, since the function $| v_1(t) - v_2(t) |$ is increasing (see (1.11)), we have

$$| u_1(t) - u_2(t) | = \lim | v_1(t + s_{\gamma'}) - v_2(t + s_{\gamma'}) | = \mathrm{const}$$

and (4.6) is proved for solutions u_1, u_2 such that $u_1(0), u_2(0) \in PK$.

Now, since the operators P_γ are noncontractive, P is noncontractive too:

$$| Py_1 - Py_2 | \geqslant | y_1 - y_2 |.$$

Moreover, it is not difficult to see that $PK \subset K$. So $PK = K$ (see [15], Ch. VI, § 11, Ex. 16) and the proof is complete. □

PROPOSITION 4.7. *The set of all bounded solutions of* (1.2) *is convex.*

Proof. Let $u_0, u_1 \in BS^p(\mathbf{R}; V) \cap C_b(\mathbf{R}; H)$ be solutions of (1.2) and $c = | u_1(0) - u_2(0) |$. By Proposition 1.1 there is a solution u_α of (1.2) on $(t_0, +\infty)$ such that

$$u_\alpha(t_0) = (1 - \alpha) u_0(t_0) + \alpha u_1(t_0),\ \alpha \in (0, 1).$$

For $t \geqslant t_0$ we have

$$| u_\alpha(t) - u_0(t) | \leqslant | u_\alpha(t_0) - u_0(t_0) | = \alpha c, \tag{4.7}$$

$$|u_\alpha(t)-u_1(t)| \leqslant |u_\alpha(t_0)-u_1(t_0)| = (1-\alpha)c. \tag{4.8}$$

Proposition 4.6 implies $|u_0(t)-u_1(t)|=c$. Taking into account the strict convexity of the norm of H, we obtain from this and from (4.7), (4.8) the equality

$$u_\alpha(t) = (1-\alpha)u_0(t)+\alpha u_1(t),\ t \geqslant t_0.$$

Since t_0 is arbitrary, we see that $(1-\alpha)u_0+\alpha u_1$ is a solution of (1.2) on the axis. □.

Now, we state

THEOREM 4.8. *Suppose that the embedding $V \subset H$ is compact and that the operators $A(t)$ satisfy condition* $(\mathrm{cap}_{V,p})$ *and inequalities* (1.3), (1.5). *Then for any $f \in S^{p'}(\mathbf{R}; V')$ there is a solution $u \in \mathrm{CAP}(\mathbf{R}; H)$ of inequality* (1.2).

Proof. Here we use the classical min-max procedure. Let

$$\mu = \inf \| u \|_{C_b(\mathbf{R};H)}, \tag{4.9}$$

where the infimum is taken over all bounded solutions of (1.2). First of all we will show that the minimum in (4.9) is achieved (the corresponding solution is called a minimal solution). Indeed, let $\{u_k\}$ be a minimizing sequence for (4.9). By Lemma 4.1 we may assume that $\lim u_k=u$ exists in $C(\mathbf{R}; H)$ and exists weakly in $L^p_{\mathrm{loc}}(\mathbf{R}; V)$. By Proposition 1.6, u is a solution of (1.2), it is clearly minimal.

Now we claim that the minimal solution is unique. Indeed, let u_1 and u_2 be two minimal solutions. By Proposition 4.7, $u=(u_1+u_2)/2$ is also a solution of (1.2) and by Proposition 4.6,

$$|u_1(t)-u_2(t)| = c.$$

Now the equality

$$|u_1+u_2|^2+|u_1-u_2|^2 = 2(|u_1|^2+|u_2|^2)$$

implies

$$|u(t)|^2+\frac{c^2}{4} \leqslant \mu^2.$$

This contradicts the minimality of u_1 and u_2, provided $u_1 \neq u_2$.

Now we will prove the almost periodicity of the minimal solution. Previous arguments apply to all inequalities $VI\ (A(\cdot+s), f(\cdot+s)), s\in\mathbf{R}_B$. This gives rise to numbers $\mu(s)$ and corresponding minimal solutions $u(\cdot, s)$. Let $s_1, s_2, \in\mathbf{R}_B$. We find a net $\{t_\gamma\}\subset\mathbf{R}$ such that $s_1+t_\gamma\to s_2$ in $\mathbf{R}_B$ (this may be done since $s_1+\mathbf{R}$ is dense in $\mathbf{R}_B$). By Lemma 4.1 we may assume that $\lim u(\cdot, s_1+t_\gamma)=v(\cdot)$ exists in $C(\mathbf{R}; H)$ and exists weakly in $L^p_{loc}(\mathbf{R}; V)$. Corollary 1.7 shows that v is a solution of inequality $VI\ (A(\cdot+s_2), f(\cdot+s_2))$. Using the identity $u(\cdot+\tau, s)=u(\cdot, s+\tau), \tau\in\mathbf{R}, s\in\mathbf{R}_B$, which follows from the uniqueness of the minimal solution, we obtain

$$\mu(s_1) = \sup_{t\in\mathbf{R}}|\, u(t, s_1)\,| = \sup_{t\in\mathbf{R}}|\, u(t+t_\gamma, s_1)\,| =$$

$$= \sup_{t\in\mathbf{R}}|\, u(t, s_1+t_\gamma)\,| \geqslant \sup_{t\in\mathbf{R}}|\, v(t)\,| \geqslant \mu(s_2).$$

Interchanging s_1 and s_2 we see that $\mu(s_1)=\mu(s_2)$, and that v is the minimal solution of $VI\ (A(\cdot+s_2), f(\cdot+s_2))$, i.e. $v(\cdot)=u(\cdot, s_2)$. Hence $u(\cdot, s)$ is a continuous function of $s\in\mathbf{R}_B$ with values in $C(\mathbf{R}; H)$. Setting $\tilde{u}(s)=u(0, s)$, we obtain a continuous extension, $\tilde{u}$, of the minimal solution u of (1.2) to $\mathbf{R}_B$. So the almost periodicity of u is proved. □

As for uniqueness of the solution in the case under consideration, we have

PROPOSITION 4.9. *Under the conditions of the Theorem* 4.8, *assume that there is a set* $M\subset\mathbf{R}$ *of positive Lebesgue measure such that* $A(t)$ *is a strictly monotone operator for any* $t\in M$. *Then inequality* (1.2) *has a unique solution* $u\in C_b(\mathbf{R}; H)$.

Proof. Let u_1 and u_2 be two bounded solutions of (1.2). The invariance identity and inequality (1.10) imply

$$\int_{t_1}^{t_2}(A(t)u_1(t)-A(t)u_2(t),\ u_1(t)-u_2(t))dt = 0,\ t_1 \leqslant t_2.$$

From the strict monotonicity we deduce from this that u_1 and u_2 coincide almost everywhere on M. Now, the invariance identity shows that $u_1 \equiv u_2$. □.

REMARK 4.10. In the proof of Theorem 4.8 coerciveness (i.e. inequality (1.5)) was used at the following two points:

1) to prove Proposition 4.7 we need an existence theorem for solutions of the Cauchy problem for variational inequalities;

2) to guarantee existence of at least one solution in $BS^p(\mathbf{R}; V) \cap C_b(\mathbf{R}; H)$ we use Theorem 4.3.

However, there are existence theorems for solutions of the Cauchy problem under somewhat weaker assumptions than (1.5) [21, 118]. Under the conditions of those theorems Proposition 4.7 remains valid. Therefore, if for some given $f \in S^{p'}(\mathbf{R}; V')$ the existence of a bounded solution is known, then our previous arguments (with Lemma 4.1 replaced by Remark 4.2) permit us to deduce existence of a solution in the space $CAP(\mathbf{R}; H)$. In fact, it is sufficient to assume only the existence of a solution $u_+ \in BS^p(\mathbf{R}_+; V) \cap C_b(\mathbf{R}_+; H)$. Indeed, we can find a net $\{t_\gamma\} \subset \mathbf{R}$ such that $t_\gamma \to +\infty$ in $\mathbf{R}$ and $t_\gamma => 0$ in $\mathbf{R}_B$. The functions $u_+(\cdot + t_\gamma)$ are solutions of the inequalities $VI\ (A(\cdot + t_\gamma), f(\cdot + t_\gamma))$ on $[-t_\gamma, +\infty)$. According to Remark 4.2, we may assume that $u_+(\cdot + t_\gamma) \to u(\cdot)$ in $C([a, b]; H)$ and weakly in $L^p(a, b; V)$ for any $a < b$. Then $u(\cdot)$ is a solution of $VI\ (A, f)$ and it clearly belongs to the space $BS^p(\mathbf{R}; V) \cap C_b(\mathbf{R}; H)$.

REMARK 4.11. In the situation of Remark 1.3 there is a statement similar to Theorem 4.8.

5. Almost periodicity in the sense of Besicovitch.

5.1. SETTING OF THE PROBLEM.

First of all we need to weaken even further the formulation of the problem of variational inequalities, so that it becomes meaningful for Besicovitch a.p. functions. To this end we discuss the operator of differentiation on spaces of such functions and obtain some auxiliary results.

Let $E \subset E'$ be reflexive Banach spaces. We set

$$\mathfrak{D}(\mathbf{R}_B; E) = CAP^\infty(\mathbf{R}; E) = \tag{5.1}$$

$$\{f \in CAP(\mathbf{R}; E) \mid f^{(n)} \in CAP(\mathbf{R}; E),\ n \in \mathbf{Z}_+\}$$

This is a Fréchet space in the natural topology. We denote by $\mathfrak{D}'(\mathbf{R}_B; E')$ the space dual to $\mathfrak{D}(\mathbf{R}_B; E)$ (its elements may be naturally considered as Besicovitch distributions). The symbol $(\cdot, \cdot)_B$ stands for duality between the spaces $\mathfrak{D}'(\mathbf{R}_B; E')$ and $\mathfrak{D}(\mathbf{R}_B; E)$. The space $B^1(\mathbf{R}; E')$ is embedded into $\mathfrak{D}'(\mathbf{R}_B; E')$ by the map

$$v \mapsto (v, f)_B = \mathbf{M}\{(v, f)\},\ f \in \mathfrak{D}(\mathbf{R}_B; E). \tag{5.2}$$

So, the duality $(\cdot,\cdot)_B$ just defined is compatible with that of Ch. 1. We define the operator of differentiation on $\mathfrak{D}'(\mathbf{R}_B;E')$ by the formula

$$(v',f)_B = -(v,f')_B,\ v\in\mathfrak{D}'(\mathbf{R}_B;E'),\ f\in\mathfrak{D}(\mathbf{R}_B;E'), \tag{5.3}$$

The restriction of the map (5.2) to $\mathfrak{D}(\mathbf{R}_B, E')$ embeds this space into $\mathfrak{D}(\mathbf{R}_B, E')$ and the operator of differentiation, defined by (5.2), clearly coincides on $\mathfrak{D}(\mathbf{R}_B, E')$ with the usual derivative (compare with Proposition 6.2 of Ch. 1). In particular, if V is an intermediate space between E and E' (i.e. $E\subset V\subset E'$), then for $v\in B^r(\mathbf{R};E)$ the derivative $v'\in\mathfrak{D}'(\mathbf{R}_B;E')$ is defined and we may require it to belong to a space $B^q(\mathbf{R};V)$ or $B^q(\mathbf{R};V')$ (for example).

Below the following situation will occur. Let $\mathfrak{E}$ be a locally convex Hausdorff space and V a reflexive Banach space such that $\mathfrak{E}\subset V\subset\mathfrak{E}'$ and $\mathfrak{E}\subset V'\subset\mathfrak{E}'$ (all embeddings are continuous and dense). In addition we assume that the embedding $\mathfrak{E}\subset V\cap V'$ is dense. Then we may apply the previous constructions to $E=V\cap V'$, and define weak derivatives of a.p. functions with values in any space intermediate between $V\cap V'$ and $V+V'$.

We return now to the situation described in n°. 3.3. Let V be a reflexive Banach space, H a Hilbert space and $V\subset H\subset V'$. Also, let $A(t)\colon V\to V'$, $t\in\mathbf{R}$, be a family of monotone semicontinuous operators satisfying condition $(\mathrm{cap}_{V,p})$ and inequalities (1.3) and (1.5), and let ϕ be a proper convex lower semicontinuous function on V. As usual we assume without loss of generality that $\phi\geqslant 0$. For $v\in B^p(\mathbf{R};V)$ we set

$$\Phi(v) = \begin{cases}\mathbf{M}\{\phi(v)\}, & \phi(v)\in B^1(\mathbf{R}),\\ +\infty, & \phi(v)\notin B^1(\mathbf{R}).\end{cases}$$

By remark 2.9 of Ch. 2, Φ is a proper convex lower semicontinuous function on $B^p(\mathbf{R};V)$. Further, by the results of n°. 2.2 (Ch. 2) the formula $(Av)(\cdot)=A(\cdot)v(\cdot)$ defines a bounded monotone semicontinuous operator

$$A\colon B^p(\mathbf{R};V)\to B^{p'}(\mathbf{R};V')$$

satisfying the inequality

$$(Av,v)_B \geqslant \alpha\,||\,v\,||^p_{B^p}+\beta,\ v\in B^p(\mathbf{R};V).$$

Now we consider the Bochner-Fejér kernels $\{\theta_\gamma\}$. They are assumed to be even (see Remark 2.5 of Ch. 1). We set

$$v_\gamma = L_\gamma v = \theta_\gamma \star_B v.$$

LEMMA 5.1. *For* $v \in B^p(\mathbf{R}; v)$ *the following statements hold:*

$$\lim \Phi(v_\gamma) = \Phi(v), \tag{5.4}$$

$$(v'_\gamma, v_\gamma - v) = 0. \tag{5.5}$$

Proof. Since $v = \lim v_\gamma$ in $B^p(\mathbf{R}; V)$ and Φ is lower semicontinuous, we have

$$\liminf \Phi(v_\gamma) \geqslant \Phi(v).$$

But $\theta_\gamma \geqslant 0$ and $\mathbf{M}\{\theta_\gamma\} = 1$. So, by the convexity of ϕ we obtain

$$\phi(\theta_\gamma \star_B v) \leqslant \theta_\gamma \star_B \phi(v).$$

Therefore $\Phi(v_\gamma) \leqslant \Phi(v)$ and

$$\limsup \Phi(v_\gamma) \leqslant \Phi(v).$$

This implies (5.4).

To prove (5.5) we may assume that $v \in \mathrm{Trig}(\mathbf{R}; V)$. We have

$$v'_\gamma = \theta'_\gamma \star_B v = \theta_\gamma v.$$

The real-valued functions θ_γ and θ'_γ are even and odd, respectively. So L_γ is a symmetric operator, while θ_γ is a skew-symmetric operator. Moreover, they commute. Now we have

$$(v'_\gamma, v_\gamma - v)_B = (\theta_\gamma v, L_\gamma v - v)_B = -(v, \theta_\gamma(L_\gamma - I)v)_B. \tag{5.6}$$

It is clear that $\theta_\gamma(L_\gamma - I)$ is skew-symmetric and we obtain (5.5). □

PROPOSITION 5.2. *Let* $f \in S^{p'}(\mathbf{R}; V')$. *Then for any solution* $u \in \mathrm{CAP}(\mathbf{R}; H) \cap \mathrm{BS}^p(\mathbf{R}; V)$ *of inequality* (1.2) *we have*

$$\mathbf{M}\{(v' + Au - f, v - u) + \phi(v) - \phi(u)\} \geqslant 0, \tag{5.7}$$

$\forall v \in B^p(\mathbf{R}; V), v' \in B^{p'}(\mathbf{R}; V')$.

Proof. First we assume that in addition $v \in S^{p'}(\mathbf{R}; V) \cap \mathrm{CAP}(\mathbf{R}; H)$ and $v' \in S^{p'}(\mathbf{R}; V')$. We assume also that all a.p. functions under consideration are extended to $\mathbf{R}_B$. Then for $s \in \mathbf{R}_B$ we have

$$v(\cdot + s) \in C(\mathbf{R}_B; \mathrm{BS}^p(\mathbf{R}; V) \cap L^\infty(\mathbf{R}; H));$$

$$v'(\cdot + s), f(\cdot + s) \in C(\mathbf{R}_B; \mathrm{BS}^{p'}(\mathbf{R}; V'));$$

$u(t+s)$ is measurable on $\mathbf{R} \times \mathbf{R}_B$ and $u(\cdot + s) \in \mathrm{BS}^p(\mathbf{R}; V)$ for almost all $s \in \mathbf{R}_B$. (this follows from Theorem 5.14 of Ch. 1 and the fact that, by Remark 3.10, u is a weakly S^p-a.p. function); $(\mathbf{A}u)(\cdot + s) = A(\cdot + s)u(\cdot + s) \in \mathrm{BS}^{p'}(\mathbf{R}; V')$ for almost all $s \in \mathbf{R}_B$. Moreover, by Proposition 2.8 of Ch. 2, the function $\phi(u(t+s))$ is measurable on $\mathbf{R} \times \mathbf{R}_B$.

Now we have

$$\int_{\mathbf{R}_B} d\mu(s) \int_0^1 \phi(u(s+t))dt = \int_0^1 dt \int_{\mathbf{R}_B} \phi(u(s+t))d\mu(s) = \int_{\mathbf{R}_B} \phi(u(s))d\mu(s). \tag{5.8}$$

However, $u(\cdot + s)$ is a solution of the shifted inequality VI $(A(\cdot + s), f(\cdot + s))$ (see (3.24)). Hence

$$\int_0^1 \phi(u(s+t))dt \leqslant c,$$

where $c > 0$ is independent of $s \in \mathbf{R}_B$. By (5.8),

$$\int_{\mathbf{R}_B} \phi(u(s))d\mu(s) \leqslant c < \infty,$$

and $\phi(u) \in B^1(\mathbf{R})$. Now we have the inequality (see (3.24) for $t_1 = 0$ and $t_1 = 1$)

$$\int_0^1 [(v'(s+t) + A(s+t)u(s+t) - f(s+t), v(s+t) - u(s+t)) +$$

$$+ \phi(v(s+t)) - \phi(u(s+t))]dt \geqslant \frac{1}{2} | v(s+t) - u(s+t) |^2 |_0^1. \tag{5.9}$$

Integrating this with respect to $s \in \mathbf{R}_B$, changing the order of integration, by Fubini's theorem, and using the fact that the measure μ is translation invariant, we obtain

$$\int_{\mathbf{R}_B} [(v' + \mathbf{A}u - f, v - u) + \phi(v) - \phi(u)] d\mu(s) \geqslant$$

$$\geqslant \frac{1}{2} \int_{\mathbf{R}_B} | v(s+1) - u(s+1) |^2 d\mu(s) - \frac{1}{2} \int_{\mathbf{R}_B} | v(s) - u(s) |^2 d\mu(s) = 0.$$

This coincides with (5.7).

We must get rid of the additional conditions imposed on v. If $v \in B^p(\mathbf{R}; V)$ and $v' \in B^{p'}(\mathbf{R}; V')$, then $v_\gamma = L_\gamma v \in \mathrm{CAP}^\infty(\mathbf{R}; V)$, $\lim v_\gamma = v$ in $B^p(\mathbf{R}; V)$ and $\lim v'_\gamma = v'$ in $B^{p'}(\mathbf{R}; V')$. We now substitute v_γ for v in (5.7) (which may be done, according to what has been said above). By (5.4),

$$\lim \mathbf{M}\{\phi(v_\gamma)\} = \mathbf{M}\{\phi(v)\}$$

and by passing to the limit we obtain the required. □.

REMARK 5.3. The statement of Proposition 5.2 can be extended in an obvious way to the situation described in Remark 1.3 (see also Theorem 2.9 and Remark 3.13). In particular, in the important case, encountered below, when A is replaced by $A + \lambda I$, $\lambda > 0$, inequality (5.7) must be satisfied for $v \in B^p(\mathbf{R}; V) \cap B^2(\mathbf{R}; H)$ and $v' \in B^{p'}(\mathbf{R}; V')$.

Now we arrive at the required setting of the problem: Given $f \in B^{p'}(\mathbf{R}; V')$, find $u \in B^p(\mathbf{R}; V)$ such that

$$(v' + \mathbf{A}u - f, v - u)_B + \Phi(v) - \Phi(u) \geqslant 0,$$

$$\forall v \in B^p(\mathbf{R}; V),\ v' \in B^{p'}(\mathbf{RT}; V'). \tag{5.11}$$

5.2. EXISTENCE OF SOLUTIONS.

First, let $f \in B^{p'}(\mathbf{R}; V')$. We consider the solution $u_\lambda \in BS^p(\mathbf{R}; V) \cap C_b(\mathbf{R}; H)$ of the following auxiliary inequality

$$\int_{t_1}^{t_2} [v' + Au_\lambda + \lambda u_\lambda - f, v - u_\lambda) + \phi(v) - \phi(u_\lambda)]dt \geq \frac{1}{2} | v - u_\lambda |^2 |_{t_1}^{t_2} \tag{5.12}$$

for any test function v. Such a solution exists and is unique by Theorem 2.5 (or Theorem 2.9, if $1<p<2$). By Theorem 3.9, $u \in CAP(\mathbf{R}; H)$ and u_λ is weakly S^p-a.p. with values in V. Hence, by Theorem 5.14 of Ch. 1, $u_\lambda \in B^p(\mathbf{R}; V)$.

PROPOSITION 5.4. *There is a subsequence $\lambda_j \to 0$ such that $u_{\lambda_j} \to u$ weakly in $B^p(\mathbf{R}; V)$ and $\star$-weakly in $B^\infty(\mathbf{R}; H)$. In addition, u is a solution of (5.11) and*

$$|| u ||_{B^p} \leq c || f ||_{B^{p'}}^{1/(p-1)},$$

$$\Phi(u) \leq c || f ||_{B^{p'}}^{p'}.$$

Proof. The solution u_λ satisfies (1.7) with u replaced by u_λ and the constant α does not depend on λ. A similar inequality also holds for the functions with arguments shifted over $s \in \mathbf{R}_B$:

$$\frac{1}{2} | u_\lambda(t+s) |^2 |_{t_1}^{t_2} + \alpha \int_{t_1}^{t_2} || u_\lambda(t+s) ||^p dt + \int_{t_1}^{t_2} \phi(u_\lambda(t+s))dt \leq$$

$$\leq \int_{t_1}^{t_2} (f(t+s), u_\lambda(t+s))dt.$$

Integrating with respect to $s \in \mathbf{R}_B$ for $t_1 = 0$ and $t_2 = 1$ we obtain (just as in the derivation of (5.10)):

$$|| u_\lambda ||_{B^p} \leq c || f ||_{B^{p'}}^{1/(p-1)}, \tag{5.13}$$

$$|| \phi(u_\lambda) ||_{B^1} \leq c || f ||_{B^{p'}}^{p'}, \tag{5.14}$$

where $c>0$ is independent of f and λ. Moreover, by Remark 2.7, $\{u_\lambda\}$ is bounded in $\mathrm{CAP}(\mathbf{R};H)$, hence also in $B^\infty(\mathbf{R};H)$. Since V is reflexive and $1<p<\infty$, $B^p(\mathbf{R};V)$ is reflexive, and $B^\infty(\mathbf{R};H)$ is dual to $B^1(\mathbf{R};H)$. Consequently, $\{u_\lambda\}$ is weakly compact in $B^p(\mathbf{R};V)$ and $\star$-weakly compact in $B^\infty(\mathbf{R};H)$. This implies that there is a subsequence $\{u_{\lambda_j}\}$ such that $u_{\lambda_j}\to u$ in the required sense. In addition, by the boundedness of $\{Au_\lambda\}$ in $B^{p'}(\mathbf{R};V')$ we may assume that $\mathbf{A}u_{\lambda_j}\to g$ weakly in $B^{p'}(\mathbf{R};V')$. Below we will omit the index j, for brevity.

Since Φ is lower semicontinuous.

$$\Phi(u)\leqslant \liminf\Phi(u_\lambda). \tag{5.15}$$

Proposition 5.2 (and Remark 5.3 in the case $1<p<2$) shows that u satisfies the inequality

$$(v'+\mathbf{A}u_\lambda+\lambda u_\lambda-f,v-u_\lambda)_B+\Phi(v)-\Phi(u_\lambda)\geqslant 0,$$

$$\forall v\in B^p(\mathbf{R};V)\cap B^2(\mathbf{R};H),\ v'\in B^{p'}(\mathbf{R};V'). \tag{5.16}$$

To pass to the limit in the last inequality we will show that

$$\limsup[(v'+\mathbf{A}u_\lambda+\lambda u_\lambda-f,v-u_\lambda)_B+\Phi(v)-\Phi(u_\lambda)]\leqslant$$

$$\leqslant (v'+\mathbf{A}u-f,v-u)_B+\Phi(v)-\Phi(u). \tag{5.17}$$

It is obvious that

$$\lim(v'-f,v-u_\lambda)_B=(v'-f,v-u_\lambda)_B. \tag{5.18}$$

Moreover,

$$\lim(\lambda u_\lambda,v-u_\lambda)_B=0, \tag{5.19}$$

since $\{u_\lambda\}$ is bounded in $B^2(\mathbf{R};H)$.

Now we will show that

$$\limsup(\mathbf{A}u_\lambda,v-u_\lambda)_B\leqslant(\mathbf{A}u,v-u_\lambda)_B. \tag{5.20}$$

To prove this inequality, we note that (5.16) together with (5.15), (5.18) and (5.19) implies the inequality

$$\limsup(\mathrm{A}u_\lambda, u_\lambda)_B \leqslant (v'-f, v-u)_B+(g, v)_B+\Phi(v)-\Phi(u),$$

where v is an arbitrary test function as in (516). From this we obtain

$$\limsup(\mathrm{A}u_\lambda, u_\lambda-u)_B \leqslant (g-f, v-u)_B+(v', v-u)_B+\Phi(v)-\Phi(u). \tag{5.21}$$

Setting $v=u_\gamma=L_\gamma u$ in the right- hand side of (5.21) and passing to the limit with respect to γ by Lemma 5.1, we obtain

$$\limsup(\mathrm{A}u_\lambda, u_\lambda-u)_B \leqslant 0. \tag{5.22}$$

Since $\mathrm{A}: B^p(\mathbf{R}; V)\to B^{p'}(\mathbf{R}; V')$ is a monotone bounded semicontinuous operator, it is also pseudomonotone (see [44], Proposition 2.5 of Ch. 2). Therefore, (5.22) implies (5.20).

Now (5.17) is an immediate consequence of (5.15) and (5.18)-(5.20). Inequality (5.17) implies (5.11) for any v, as in (5.16). Now, if $v\in B^p(\mathbf{R}; V)$, $v'\in B^{p'}(\mathbf{R}; V)$ then $v_\gamma\in \mathrm{CAP}^\infty(\mathbf{R}; V)$, $\lim v_\gamma=v'$ in $B^p(\mathbf{R}; V)$, $\lim v'_\gamma=v'$ in $B^{p'}(\mathbf{R}; V')$ and $\lim\Phi(v_\gamma)=\Phi(v)$. This permits us to say that u satisfies (5.11) for all $v\in B^p(\mathbf{R}; V)$ such that $v'\in B^{p'}(\mathbf{R}; V')$. The estimates for u follows from (5.13) and (5.14). □.

Now we state

THEOREM 5.5. *Suppose that the family of operators $A(t)$ satisfies condition* ($\mathrm{cap}_{V,p}$) *and inequalities* (1.3), (1.5). *Then for any $f\in B^{p'}(\mathbf{R}; V')$ there is a solution $u\in B^p(\mathbf{R}; V)$ of problem* (5.11). *Any two solutions u_1 and u_2 of* (5.11) *with $f=f_i$, $i=1, 2$, satisfy the estimate*

$$(\mathrm{A}u_1-\mathrm{A}u_2, u_1-u_2)_B \leqslant (f_1-f_2, u_1-u_2)_B. \tag{5.23}$$

In particular, if for all $t\in\mathbf{R}$,

$$(A(t)v-A(t)w, v-w) \geqslant c(v, w) \geqslant 0, \tag{5.24}$$

where $c(v, w)\geqslant 0$ does not depend on t and $c(v, w)\neq 0$ for $v\neq w$, then a solution of (5.11) is unique for any $f\in B^{p'}(\mathbf{R}; V')$.

Proof. Let $f_n\in S^{p'}(\mathbf{R}; V')$ and $\lim f_n=f$ in $B^{p'}(\mathbf{R}; V')$. By Proposition 5.4, there are solutions $u_n\in B^p(\mathbf{R}; V)$ of (5.11) with right-hand sides f_n such that $\|u_n\|_{B^p}\leqslant c$ and $\Phi(u_n)\leqslant c$. Passing to a subsequence, if necessary, we may assume that $u_n\to u$

weakly in $B^p(\mathbf{R}; V)$. It now remains to pass to the limit in the inequality

$$(v' + \mathbf{A}u_n - f_n, v - u_n)_B + \Phi(v) - \Phi(u_n) \geqslant 0,$$

which is done just as at the end of the proof of Proposition 5.4 (using Lemma 5.1 and pseudomonotonicity of $\mathbf{A}$).

To prove (5.23) it suffices to set $v = v_\gamma = \theta_\gamma \star_B ((u_1 + u_2)/2)$ in the inequalities

$$(v' + \mathbf{A}u_i - f_i, v - u_i)_B + \Phi(v) - \Phi(u_i) \geqslant 0, i = 1, 2,$$

add them and pass to the limit by using Lemma 5.1. □.

REMARK 5.6. In the case $f \in S^{p'}(\mathbf{R}; V')$, Proposition 5.4 yields more exact information about the solution; namely, $u \in B^\infty(\mathbf{R}; H)$. We also note that $u(\cdot + s)$ belongs to $BS^p(\mathbf{R}; V) \cap L^\infty(\mathbf{R}; H)$ for almost all $s \in \mathbf{R}_B$. However, it is not clear whether this function satisfies the inequalities $VI\ (A(\cdot + s), f(\cdot + s))$.

REMARK 5.7. The above results can be generalized to the case when $V = \bigcap V_i$ and $A(t) = \sum A_i(t)$, where $A_i(t): V_i \to V_i'$, $i = 1, \ldots, n$, are operators with suitable properties (see Remark 1.3 and Theorem 2.9).

REMARK 5.8. The restriction $V \subset H \subset V'$ does not appear explicitly in the statement of the problem (5.11). In fact we may consider the following situation. Let $\mathfrak{G}$ be a locally convex Hausdorff space and V a reflexive Banach space such that $\mathfrak{G} \subset V \subset \mathfrak{G}'$ and $\mathfrak{G} \subset V' \subset \mathfrak{G}'$ (all embeddings are continuous and dense). Assume that $(u, v)_{\mathfrak{G}} = (u, v)$ for $u, v \in \mathfrak{G}$, where $(\cdot, \cdot)_{\mathfrak{G}}$ and $(\cdot, \cdot)$ are canonical bilinear forms on $\mathfrak{G}' \times \mathfrak{G}$ and $V' \times V$, respectively. In addition we assume that the embedding $\mathfrak{G} \subset V \cap V'$ is dense. The present situation is sufficiently typical (see, for example, [44], Ch. 2, § 7). Under these assumptions the statement of problem (5.11) make sense. It is not difficult to see that the weak derivative d/dt (see n°. 5.1), viewed as an unbounded operator from $B^p(\mathbf{R}; V)$ into $B^{p'}(\mathbf{R}; V')$, is a maximal monotone operator (i.e. it and its adjoint operator are $\geqslant 0$). Now a general result of H. Brezis (see [118], Theorem II. 1) and Propositions 2.5 and 2.8 of Ch. 2 imply that for any $f \in B^{p'}(\mathbf{R}; V')$ there is a solution $u \in B^p(\mathbf{R}; V)$ of (5.11), provided the other assumptions of Theorem 5.5 are valid. The question on the validity of the estimate (5.23) and on uniqueness of a solution (under assumption (5.24)) is still open.

5.3. THE INVERSE OPERATOR AND REGULARITY OF SOLUTIONS.

As in n°. 3.1 we consider the operator $F_g: h \to u$, where u is a solution of (5.11) with $f = g + h$. Here we place restrictions on the operators $A(t)$ which ensure that (5.24) holds and, by the same token, that the solution is unique.

THEOREM 5.9. *Suppose that the conditions of Theorem 5.5 are satisfied, that* (3.1) *or* (3.2) *with* $q \geqslant 2$ *holds, and that* $g \in B^{p'}(\mathbf{R}; V')$. *Then* $F_g: B^{\rho'}(\mathbf{R}; E') \to B^{\rho}(\mathbf{R}; E)$, *where* $\rho = \min(p, q)$, *is a Hölder operator with exponent* $(q-1)^{-1}$ *on every ball in the space* $B^{\rho'}(\mathbf{R}; E')$ *(on the whole space, if* $q \leqslant p$*).*

Proof. Let $h_i \in B^{\rho'}(\mathbf{R}; E')$ and $u_i = F_g(h_i)$, $i = 1, 2$. Integrating (3.1) or (3.2) on $\mathbf{R}_B$ and using in the second case Hölder's inequality with exponents p/q and $p/(p-q)$, we deduce from (5.23) the inequality

$$||k_1 - k_2||_{B^{\rho'}} \cdot ||u_1 - u_2||_{B^\rho} \geqslant$$

$$\geqslant \begin{cases} \alpha ||u_1 - u_2||^q_{B^\rho}, & q \leqslant p (\rho = q), \\ \alpha ||u_1 - u_2||^q_{B^\rho}(1 + ||u_1||_{B^\rho} + ||u_2||_{B^\rho})^{p-q}, & p < q (\rho = p). \end{cases}$$

From this we derive the required result. □.

As in § 3 we obtain the following statement.

COROLLARY 5.10. *Let* X *be a topological space and* $A(t, x)$, $x \in X$, $t \in \mathbf{R}$, *a family of monotone semicontinuous operators that for every* x *satisfy the assumptions of Theorem 5.9 with uniform estimates with respect to* x. *Suppose that* $\lim_{x \to x_0} d_{E,\rho}(x, x_0) = 0$ *(see* (3.8)) *and that* $f(x) = g + h(x)$, *where* $g \in B^{p'}(\mathbf{R}; V')$, $h(x) \in B^{\rho'}(\mathbf{R}; E)$, *and* $\lim_{x \to x_0} h(x) = h(x_0) = 0$ *in this space. Then the unique solution* $u(x)$ *of problem* (5.11) *with a parameter satisfies* $\lim_{x \to x_0} u(x) = u(x_0)$ *in* $B^\rho(\mathbf{R}; E)$. *Moreover,*

$$||u(x) - u(x_0)||_{B^\rho} \leqslant C(||f(x) - f(x_0)||^{1/(q-1)}_{B^{\rho'}} + d_{E,\rho}(x, x_0)^{1/(q-1)}).$$

REMARK 5.11. Unfortunately, a suitable analogue to Corollary 3.3 can not be obtained in this situation. In its place we may consider the following result. Under the conditions of Theorem 5.5 with uniform estimates with respect to x, let $\lim_{x \to x_0} d_{V,p}(x, x_0) = 0$ and $\lim_{x \to x_0} f(x) = f(x_0)$ in $B^{p'}(\mathbf{R}; V')$. The set $\{u(x)\}$ is bounded in $B^p(\mathbf{R}; V)$ (due to the possible nonuniqueness, here $u(x)$ is any of the solutions). Then for any net $x_\gamma \to x_0$ we can find a subnet $\{x_{\gamma'}\}$ such that $u(x_{\gamma'})$

converges weakly to some solution $u_0 \in B^p(\mathbf{R}; V)$ corresponding to $x = x_0$. If, in particular, (5.24) holds at x_0, then $\lim_{x \to x_0} u(x) = u(x_0) = u_0$ weakly.

We now turn to the regularity of solutions. To this end we introduce the spaces

$$B^{p,1}(\mathbf{R}; E) = \{v \in B^p(\mathbf{R}; E) \mid v' \in B^p(\mathbf{R}; E)\},$$

$$B_r^{p,\theta}(\mathbf{R}; E) = (B^{p,1}(\mathbf{R}; E),\ B^p(\mathbf{R}; E))_{\theta,r},$$

where $1 < p < \infty$, $1 \leqslant r \leqslant \infty$, $0 < \theta < 1$ (if $r = \infty$, then $\theta = 0$ and $\theta = 1$ are also possible). Here E is a reflexive Banach space.

LEMMA 5.12. *$B^{p,1}(\mathbf{R}; E) = D(\Lambda)$, where $-\Lambda$ is the generator of the group $G(s)$ of right shifts in $B^p(\mathbf{R}; E)$.*

Proof. The group $\mathbf{R}$ acts continuously on $\mathbf{R}_B$ by shifts. This implies that $G(s)$ is strongly continuous. Therefore Λ is a closed operator. Moreover, $\mathrm{CAP}^\infty(\mathbf{R}; E) \subset D(\Lambda)$ and Λ is a closed extension of the operator of differentiation, with $D(\Lambda_0) = \mathrm{CAP}^\infty(\mathbf{R}; E)$. The operator d/dt (the weak derivative, see n°. 5.1) is the maximal closed extension of Λ_0. Clearly, $D(d/dt) = B^{p,1}(\mathbf{R}; E)$. However, d/dt commutes with the Bochner-Fejér operators. Hence it is also the minimal closed extension of Λ_0. Consequently, $D(\Lambda) = D(d/dt)$. □.

Lemma 5.12. implies that the norm in $B_r^{p,\theta}(\mathbf{R}; E)$ is given by formulas (5.12). Since E is reflexive and $p > 1$, the space $B^p(\mathbf{R}; E)$ is also reflexive. Hence (see, for example, [3], § 6.7),

$$B^{p,1}(\mathbf{R}; E) = B_\infty^{p,1}(\mathbf{R}; E)$$

It is not difficult to see that $B^{p,1}(\mathbf{R}; E)$ is reflexive. Hence (see [3], § 3.7) $B_r^{p,\theta}(\mathbf{R}; E)$, $0 < \theta < 1$, $1 < r < < \infty$, is also reflexive. Moreover, $B^{p,1}(\mathbf{R}; E)$ is dense in $B_r^{p,\theta}(\mathbf{R}; E)$ for $r < \infty$. Finally we note that there is a natural continuous embedding

$$S_r^{p,\theta}(\mathbf{R}; E) \subset B_r^{p,\theta}(\mathbf{R}; E). \tag{5.25}$$

THEOREM 5.13. *Suppose that the conditions of Theorem 5.9 are satisfied and that the estimate (3.13) holds with $0 < \gamma < 1$. Let $f \in B_r^{p',\theta}(\mathbf{R}; E')$, where $1 \leqslant r \leqslant \infty$, $0 < \theta < \gamma$ for $r < \infty$ and $0 < \theta \leqslant \gamma$ for $r = \infty$. Then the unique solution of (5.11) belongs to the space $B_\nu^{p,\mu}(\mathbf{R}; E)$, where $\nu = r(q-1)$ and $\mu = \theta/(q-1)$.*

The *Proof* is similar to that of Theorem 3.5, using Theorem 5.9 instead of Theorem 3.1. □.

COROLLARY 5.14. *Let* $q=2, \gamma=1$ *and* $f \in B^{p',1}(\mathbf{R}; E)$ *in the conditions of Theorem* 5.13. *Then the unique solution* $u \in B^p(\mathbf{R}; E)$ *of* (5.11) *satisfies* $u' \in B^p(\mathbf{R}; E)$, *and*

$$(u' + Au - f, v - u)_B + \Phi(v) - \Phi(u) \geqslant 0, \forall v \in B^p(\mathbf{R}; V). \tag{5.26}$$

REMARK 5.15. In spite of the formal analogy between these results and those of n°. 3.1 and n°. 3.2, the numerical values of their parameters may differ. Thus, the Hölder exponent $(q-1)^{-1}$ of the operator

$$F_g: B^{p'}(\mathbf{R}; E') \rightarrow B^p(\mathbf{R}; E)$$

is, in general, better than the Hölder exponent $2/q(q-1)$ of its analogue

$$F_g: BS^{p'}(\mathbf{R}; E') \rightarrow BS^p(\mathbf{R}; E).$$

However, it is equal to that of the operator

$$F_g: BS^{p'}(\mathbf{R}; E') \rightarrow C_b(\mathbf{R}; H).$$

Similar remarks can also be made about the regularity theorems. For

$$f \in S_r^{p',\theta}(\mathbf{R}; E') \subset B_\nu^{p',\theta}(\mathbf{R}; E),$$

Theorem 5.13 yields

$$u \in S_l^{p,\eta}(\mathbf{R}; E) \subset B_l^{p,\eta}(\mathbf{R}; E).$$

Here

$$l = rq(q-1)/2 \geqslant \nu = r(q-1),$$

$$\eta = 2\theta/q(q-1) \leqslant \mu = \theta/(q-1);$$

consequently (see [3])

$$B_{\nu}^{p,\mu}(\mathbf{R}, E) \subset B_{\gamma}^{p,\eta}(\mathbf{R}, E).$$

However, regarding $u(t)$ as a function with values in H, we have, by Proposition 3.12,

$$u \in \mathrm{CAP}_{\nu}^{\mu}(\mathbf{R}; H) \subset B_{\nu}^{s,\mu}(\mathbf{R}, H)$$

for all $1 \leqslant s < \infty$. We note also that if $q=2, \gamma=1$ and $f, f' \in S^{p'}(\mathbf{R}; E')$, then u belongs to $S^p(\mathbf{R}; E) \bigcap \mathrm{CAP}(\mathbf{R}; H)$ and $u' \in BS^p(\mathbf{R}; E) \bigcap L^{\infty}(R; H)$. Since u satisfies (5.11), we additionally have $u' \in B^p(\mathbf{R}; E)$ (see the discussion at the end of n°. 3.3).

6. Singular perturbation.

We consider (formally) the following inequality with a small parameter $\mu>0$ in front of the derivative:

$$(\mu u'(t) + A(t)u(t) - f(t), v - u(t)) + \phi(v) - \phi(u(t)) \geqslant 0, \ \forall v \in V, \tag{6.1$_\mu$}$$

for almost all $t \in \mathbf{R}$. If $\mu = 0$, then the problem turns into the stationary inequality

$$(A(t)u(t) - f(t), v - u(t)) + \phi(v) - \phi(u(t)) \geqslant 0, \ \forall v \in V, \tag{6.1$_0$}$$

where $t \in \mathbf{R}$ plays the role of parameter. According to n°. 1.1, a weak setting of the problem corresponding to (6.1 $_\mu$) is stated in the following way. Let the conditions of n°. 1.1 (in particular, inequalities (1.3) and (1.5) and condition (c)) be satisfied. Given $f \in L_{\mathrm{loc}}^{p'}(\mathbf{R}; V')$, find a function $u = u_\mu \in L_{\mathrm{loc}}^{p}(\mathbf{R}; V) \bigcap C(\mathbf{R}; H)$ such that

$$\int_{t_1}^{t_2} [(uv' + Au - f, v - u) + \phi(v) - \phi(u)]dt \geqslant \frac{\mu}{2} | v - u |^2 |_{t_1}^{t_2}, \tag{6.2$_\mu$}$$

for any test function v. The corresponding limit inequality is

$$\int_{t_1}^{t_2} [(Au - f, v - u) + \phi(v) - \phi(u)]dt \geqslant 0, \ t_1 \leqslant t_2, \tag{6.2$_0$}$$

for any $v \in L_{\mathrm{loc}}^{p}(\mathbf{R}; V)$. This is clearly equivalent to (6.1_0). It is easy to see that under suitable conditions all previous results are applicable to (6.2_μ) for any fixed

$\mu>0$ (it is sufficient to divide (6.2_μ) by μ.

The usual stationary theory (see, for example, [44, 118]) is applicable to (6.2_0). We present here the corresponding results in a suitable form.

PROPOSITION 6.1. *Assume that condition* (c) *is satisfied and that inequalities* (1.3) *and* (1.5) *hold. If* $f\in L^{p'}_{loc}(\mathbf{R}; V')$, *then there is a solution* $u\in L^p_{loc}(\mathbf{R}; V)$ *of inequality* (6.2_0). *Moreover, if* $f\in BS^{p'}(\mathbf{R}; V')$, *then* $u\in BS^p(\mathbf{R}; V)$.

Proof. By general results [44, 118], for any fixed $N>0$ problem (6.2_0) has a solution $u=u_N\subset L^p(-N, N; V)$ and

$$||u||_{L^p(-N,N;V)} \leq C||f||_{L^{p'}(-N,N;V')}, \tag{6.3}$$

where $C(x)$, $x\geq 0$, depends only on constants involved in (1.3) and (1.5). Moreover, estimate (6.3) is valid for any solution of (6.2_0) on the interval $(-N, N)$. It is easy to see that $u_N|_{(a,b)}$ is a solution of (6.2_0) on $(a, b)\subset(-N, N)$ and is bounded in $L^p(a, b; V)$. Hence passing a subsequence, if necessary, we may assume that $\lim u_N=u_0$ exists weakly in $L^p_{loc}(\mathbf{R}; V)$. Results in [142] imply that u_0 is a solution of (6.2_0) on $\mathbf{R}$ (but it is not difficult to see this immediately). The second assertion easily follows from the estimate (6.3). □

REMARK 6.2. If all operators $A(t)$ are strictly monotone, then the solution constructed is unique. Assume now that condition ($\mathrm{cap}_{V,p}$) is satisfied and that inequality (3.1) holds with $E=V$. If $f\in S^{p'}(\mathbf{R}; V')$, then we may consider a family of inequalities,

$$\int_0^1[(A(s+t)u_s(t)-f(s+t), u(t)-u_s(t))+\phi(v(t))-\phi(u_s(t))]dt \geq 0,$$

where $s\in\mathbf{R}_B$. Any such inequality has a unique solution $u_s\in L^p(0, 1; V)$, and

$$u_s(t) = u_0(s+t),\ s\in\mathbf{R},\ t\in(0, 1).$$

Results in [142] imply that $u_s(\cdot)\in L^p(0, 1; V)$ depends continuously on $s\in\mathbf{R}_B$ and, consequently, $u_0\in S^p(\mathbf{R}; V)$. Using results in [142] it can also be shown that $f\in CAP(\mathbf{R}; V)$ implies $u_0\in CAP(\mathbf{R}; V)$.

Now we consider the behaviour of bounded solutions, u_μ, of (6.2_μ) as $\mu\to 0$. Under the conditions of Theorem 2.7 such solutions are uniquely determined.

THEOREM 6.3. *Assume that condition* (c) *is satisfied, inequalities* (1.3) *and* (1.5) *hold, and one of inequalities* (3.1) *or* (3.2) *hold with* $q\geq 2$. *Then the family*

$\{u_\mu\}_{0<\mu\leqslant\mu_0}$ of solutions is bounded in $BS^p(\mathbf{R}; V)$ and $\lim_{\mu\to 0} u_\mu = u_0$ weakly in $L^p_{loc}(\mathbf{R}; V)$ and strongly in $L^\rho_{loc}(\mathbf{R}; E)$, where $\rho=\min(p, q)$.

Proof. Lemma 1.4 gives the inequality

$$\frac{\mu}{2}|u_{\mu n}|^2\Big|_{t_1}^{t_2} + \alpha\int_{t_1}^{t_2}||u_{\mu n}||^p dt + \int_{t_1}^{t_2}\phi(u_{\mu n})dt \leqslant \int_{t_1}^{t_2}(f_n, u_{\mu n})dt, \tag{6.4}$$

where $f_n(t)=f(t)$ for $t\geqslant -n$, $f_n(t)=0$ for $t\leqslant -n$ and $u_{\mu n}$ is the approximate solution constructed in the proof of Theorem 2.5. Setting

$$\lambda^2 = \frac{\mu}{2}|u_{\mu n}|^2,\ \psi = ||u_{\mu n}||,\ \theta = ||f_n||.$$

and disregarding the term in (6.4) containing ϕ, we obtain inequality (1.2) of Ch. 2. Note that we have the inequality

$$\lambda(t) \leqslant c\mu^{1/2}\psi(t).$$

Applying Lemma 1.1 of Ch. 2 and passing to the limit as $n\to\infty$ (see the proof of Theorem 2.5), we obtain the inequalities

$$\mu^{1/2}||u_\mu||_{C_b(\mathbf{R}; H)} \leqslant C, \tag{6.5}$$

$$||u_\mu||_{S^p} \leqslant C, \tag{6.6}$$

where $C>0$ depends only on f. So the boundedness of $\{u_\mu\}$ in the space $BS^p(\mathbf{R}; V)$ is shown. Moreover, inequalities (6.4)-(6.6) imply the boundedness of $\{\phi(u_\mu)\}$ in $BS^1(\mathbf{R})$.

Now this boundedness of $\{u_\mu\}$ implies that it is weakly precompact in $L^p_{loc}(\mathbf{R}; V)$. Let $u=\lim u_{\mu'}$ weakly in $L^p_{loc}(\mathbf{R}; V)$ as $\mu'\to 0$ along a subsequence. In addition we may assume that

$$\lim A(t)u_{\mu'}(t) = g(t)$$

weakly in $L^{p'}_{loc}(\mathbf{R}; V')$. By Proposition 2.8 of Ch. 2 we have for all $t_1\leqslant t_2$,

$$\int_{t_1}^{t_2}\phi(u)dt \leqslant \liminf\int_{t_1}^{t_2}\phi(u_\mu)dt. \tag{6.7}$$

Note that by (6.6), $\mu^{1/2}u_\mu \to 0$ in the space $BS^p(\mathbf{R}; V)$and, as a consequence, in $BS^p(\mathbf{R}; H)$. So we may assume that

$$\lim \mu^{1/2} u_\mu(t) = 0 \text{ in } H \text{ for almost all } t \in \mathbf{R}. \tag{6.8}$$

Now, passing to the limit as $\mu' \to 0$, we obtain from (6.6)-(6.8) and (6.2_μ) the inequality

$$\limsup_{\mu' \to 0} \int_{t_1}^{t_2} (A(t)u_{\mu'}(t), u_{\mu'}(t))dt \leqslant$$

$$\leqslant \int_{t_1}^{t_2} (g(t), u(t))dt - \int_{t_1}^{t_2} (f(t), v(t) - u(t))dt + \int_{t_1}^{t_2} [\phi(v) - \phi(u)]dt, \tag{6.9}$$

which holds for almost all $t_1 \leqslant t_2$ and for all test functions v. Let $\psi \in C_0^\infty(\mathbf{R})$, $\psi \geqslant 0$, $\operatorname{supp}\psi \subset (-1, -1)$, and

$$\int_{\mathbf{R}} \psi(t)dt = 1.$$

Set

$$\psi_\delta(t) = \delta^{-1}\psi(\delta^{-1}t)$$

and substitute $v = u_\delta$ into (6.9), where

$$u_\delta(t) = \int_{t_1}^{t_2} \psi_\delta(t-\tau)u(\tau)d\tau.$$

It is not difficult to see that u_δ is a test function and that $u_\delta \to u$ in the space $L^p(t_1, t_2; V)$. Since $u_\delta(t)$ is a convex combination of values of $u(\tau)$ which belong to $D(\phi)$ for almost all $\tau \in [t_1, t_2]$, and since ϕ is a convex function, we have

$$\phi(u_\delta(t)) \leqslant \int_{t_1}^{t_2} \psi_\delta(t-\tau)\phi(u(\tau))d\tau.$$

Integrating this inequality with respect to $t \in [t_1, t_2]$ and taking into account (6.10), we

obtain

$$\int_{t_1}^{t_2}\phi(u_\delta(t))dt \leqslant \int_{t_1}^{t_2}\phi(u(t))dt.$$

On the other hand, the function Φ defined by

$$\Phi(v) = \int_{t_1}^{t_2}\phi(v)dt$$

is lower semicontinuous on $L^p(t_1, t_2; V)$ (see Proposition 2.8 of Ch. 2), and so

$$\liminf_{\delta\to 0}\int_{t_1}^{t_2}\phi(u_\delta(t))dt \geqslant \int_{t_1}^{t_2}\phi(u(t))dt.$$

Hence

$$\lim_{\delta\to 0}\int_{t_1}^{t_2}\phi(u_\delta(t))dt = \int_{t_1}^{t_2}\phi(u(t))dt.$$

$$\limsup_{\mu'\to 0}\int_{t_1}^{t_2}(A(t)u_{\mu'}(t), u_{\mu'}(t))dt \leqslant \int_{t_1}^{t_2}(g(t), u(t))dt,$$

and so

$$\limsup_{\mu'\to 0}\int_{t_1}^{t_2}(A(t)u_{\mu'}(t), u_{\mu'}(t)-u(t))dt \leqslant 0.$$

Since $\mathbf{A}:L^p(t_1, t_2; V)\to L^{p'}(t_1, t_2; V)$ is a pseudomonotone operator, we have

$$\limsup_{\mu'\to 0}\int_{t_1}^{t_2}(A(t),u_{\mu'}(t), u_{\mu'}(t)-v(t))dt \leqslant \int_{t_1}^{t_2}(A(t)u(t), u(t)-v(t))dt$$

for any $v\in L^p(t_1, t_2; V)$. Taking this and (6.7), (6.8) into account and passing to the limit in (6.2_μ), we see that u satisfies inequality (6.2_0). Since u_0 is the unique solution of (6.2_0), we have $u=u_0$. Hence passage to a subsequence is superfluous and $u_\mu \to u_0$ weakly in $L^p(t_1, t_2; V)$.

Finally, inequality (6.11) and the monotonicity of $A(t)$ imply that

$$\lim_{\mu\to 0} \int_{t_1}^{t_2} (A(t)u_\mu(t) - A(t)u_0(t), u_\mu(t) - u_0(t))dt = 0, \tag{6.12}$$

and the statement concerning strong convergence follows from (3.1) or (3.2). □

REMARK 6.4. In fact, $\mu^{1/2}u_\mu \to 0$ in $C(\mathbf{R}; H)$ (not only almost everywhere). Indeed, reproducing the arguments we used to prove Lemma 4.1, and taking into account (6.6), we see that the functions $\{\mu^{1/2}u_\mu\}$, with values in H, are equicontinuous on $\mathbf{R}$. This and the convergence almost everywhere imply the required assertion.

REMARK 6.5. Under the conditions of Theorem 4.3 there is a statement similar to Theorem 6.3, but without strong convergence, provided u_μ denotes the solution constructed in § 4. Such solutions satisfy inequalities (6.5) and (6.6) and we may repeat the previous arguments. However, in view of a possible nonuniqueness, passage to a subsequence is actually needed. Hence, for a subsequence $\mu' \to 0$ we have $\lim u_{\mu'} = u_0$ weakly in $L^p_{\text{loc}}(\mathbf{R}; V)$, where u_0 is some solution of the stationary problem.

Now we turn our attention to almost periodic solutions.

THEOREM 6.6. *Let $A(t)=A_1+A_2(t)$. Assume that condition* $(\text{cap}_{E,\rho})$ *is satisfied for $A_2(t)$, that one of the inequalities* (3.1) *or* (3.2) *holds with $q \geqslant 2$, and that $f \in S^{\rho'}(\mathbf{R}; E')$, where $\rho = \min(p, q)$. Then u_0 and u_μ belong to $S^\rho(\mathbf{R}; E)$ and $\lim u_\mu = u_0$ in that space.*

Proof. The fact that $u_\mu \in S^\rho(\mathbf{R}; E)$ follows from Theorem 3.9. To prove the required convergence statement it is sufficient, according to Theorem 6.3, to prove the precompactness of $\{u_\mu\}$ in $S^\rho(\mathbf{R}; E)$. To this end we consider the functions $u_\mu^b(s) = u_\mu(s+\cdot)$ with values in $\text{BS}^\rho(\mathbf{R}; E)$ (Bochner transforms). By the almost periodicity of u_μ, they are defined and continuous at all points $s \in \mathbf{R}_B$. Theorem 6.3 may be applied to shifted inequalities

$$\int_{t_1}^{t_2} [(\mu v' + A(s+t)u - f(s+t), v-u) + \phi(v) - \phi(u)]dt \geqslant$$

$$\geqslant \frac{\mu}{2} | v-u |^2 |_{t_1}^{t_2}, \ t_1 \leqslant t_2. \tag{6.13}$$

Hence for any $s \in \mathbf{R}_B$, $\lim_{\mu->0} u_\mu^b(s)$ exists in the space $BS^p(\mathbf{R}; E)$. Therefore, by the Arzela'-Ascoli theorem, it is sufficient to prove the equicontinuity of $\{u_\mu\}$ on $\mathbf{R}_B$.

To prove this we consider the operators $F_\mu : f \to u_\mu$ corresponding to (6.2_μ) according to § 3. As in the proof of Theorem 3.1, it may be show that, uniformly in μ, the operators

$$\mu^{1/2} F_\mu : BS^{p'}(\mathbf{R}; E') \to C_b(\mathbf{R}; H),$$

$$F_\mu : BS^{p'}(\mathbf{R}; E') \to BS^p(\mathbf{R}; E)$$

are locally Hölder with exponents $1/(q-1)$ and $2/q(q-1)$, respectively (uniformly in μ means that the Hölder constant is independent of μ). At this place the character of dependence of the estimates of Lemma 1.1 (Ch. 2) on $k = c\mu^{1/2}$ play an essential role. Therefore, all statements of § 3 on continuous dependence of solutions on parameters are still valid, uniformly in μ. In particular, this holds for problem (6.13), depending on a parameter $s \in \mathbf{R}_B$. By the uniqueness, the solution of the last problem is just $u_\mu^b(s)$ and our statement is proved. □

REMARK 6.7. Under the conditions of Theorem 6.6, $\lim \mu^{1/2} u_\mu = 0$ in $CAP(\mathbf{R}; H)$. Indeed, the family of functions $\{\mu^{1/2} u_\mu^b(s)\}$ with values in $C_b(\mathbf{R}; H)$ is equicontinuous in $s \in \mathbf{R}_B$.

Now we consider the case of solutions which are a.p. in the sense of Besicovitch. Let $f \in B^{p'}(\mathbf{R}; V')$. We consider the following two problems (the notations of § 5 are used).

1. Find $u = u_\mu \in B^p(\mathbf{R}; V)$ such that

$$(uv' + \mathbf{A}u - f, v-u)_B + \Phi(v) - \Phi(u) \geqslant 0,$$

$$\forall v \in B^p(\mathbf{R}; V), \ v' \in B^{p'}(\mathbf{R}; V'). \tag{6.14$_\mu$}$$

2. Find $u = u_0 \in B^p(\mathbf{R}; V)$ such that

$$(\mathbf{A}(u)f, v-u)_B + \Phi(v) - \Phi(u) \geqslant 0, \ \forall v \in B^p(\mathbf{R}; V). \tag{6.14$_0$}$$

These problems correspond to (6.2_μ) and (6.2_0), respectively. Under the conditions of Theorem 5.5 we denote by u_μ any solution of (6.14_μ). The following statement is valid.

PROPOSITION 6.8. *Assume that the conditions of Theorem 5.5 are valid. Then the set* $\{u_\mu\}_{0<\mu\leqslant\mu_0}$ *is bounded in* $B^p(\mathbf{R}; V)$, *and for a subsequence* $\mu'\to 0$ *the limit* $\lim u_{\mu'} = u_0$ *exists weakly in that space. Moreover,* u_0 *is a solution of* (6.14_0). *If, in addition, one of the inequalities* (3.1) *or* (3.2) *holds with* $q\geqslant 2$, *then* $\lim u_{\mu'} = u_0$ *in* $B^\rho(\mathbf{R}; E)$, $\rho = \min(p, q)$.

Proof. Inequality (6.14_μ) with $v=0$ implies that $\{u_\mu\}$ is bounded. Hence we may assume that for a subsequence $\mu'\to 0$ the limit $\lim u_{\mu'} = u_0$ exists weakly in $B^p(\mathbf{R}; V)$ and that $\lim \mathbf{A}u_{\mu'} = g$ weakly in $B^{p'}(\mathbf{R}; V')$. So we need to prove that u_0 is a solution of (6.14_0). By Proposition 2.8 of Ch. 2,

$$\Phi(u_0) \leqslant \liminf \Phi(u_{\mu'}).$$

Passing to the limit in (6.14_μ) we obtain

$$\liminf (\mathbf{A}u_{\mu'}, u_{\mu'})_B \leqslant (g, v)_B - (f, v-u_0)_B + \Phi(v) - \Phi(u_0)$$

(here v must be taken as in (6.14_μ)). Now we set $v = L_\gamma u_0$, where $\{L_\gamma\}$ are the Bochner-Fejér operators. Then, passing to the limit with respect to γ and using Lemma 5.1, we obtain

$$\limsup (\mathbf{A}u_{\mu'}, u_{\mu'})_B \leqslant (g, u_0)_B.$$

As consequence,

$$\limsup (\mathbf{A}u_{\mu'}, u_{\mu'} - u_0)_B \leqslant 0,$$

and, by the pseudomonotonicity of $\mathbf{A}$,

$$\limsup (\mathbf{A}u_{\mu'}, u_{\mu'} - v)_B \leqslant (\mathbf{A}u_0, u_0 - v)_B \tag{6.15}$$

for any $v\in B^p(\mathbf{R}; V)$. Passing to the limit in (6.14_μ) and taking into account (6.15) we see that u satisfies inequality (6.14_0). Moreover, (6.15) and the monotonicity of $\mathbf{A}$ imply that

$$\lim (\mathbf{A}u_\mu - \mathbf{A}u_0, u_\mu - u_0)_B = 0.$$

This gives rise to the last statement. □

7. Some examples and additional results.

7.1. PARABOLIC INEQUALITIES.

For simplicity we consider operators which do not depend on t. For the operator A we can take any elliptic operator of order $2m$ in divergence form,

$$A(u) = \sum_{|\alpha|\leqslant m} (-1)^{|\alpha|}\partial^\alpha A_\alpha(x, \delta^m u), \tag{7.1}$$

in a smooth and bounded domain. Here $\partial=(\partial/\partial x_1, \ldots, \partial/\partial x_n)$, $\partial^\alpha=\partial^{|\alpha|}/\partial x_1^{\alpha_1}\cdots\partial x_n^{\alpha_n}$ for $\alpha=(\alpha_1, \ldots, \alpha_n)\in\mathbf{Z}_+^n$, $|\alpha|=\sum\alpha_j$, $\partial^k=\{\partial^\alpha \mid |\alpha|=k\}$, $\delta^m u=\{u, \partial u, \ldots, \partial^m u\}$.

A typical choice of spaces in this situation is as follows. Let $H=L^2(\Omega)$ and let $V\subset W^{m,p}(\Omega)$ be a closed subspace such that $W_0^{1,p}(\Omega)\subset V$. Under reasonable assumptions on A_α the operator (7.1) maps V into V' continuously, and inequality (1.3) holds (see, for example, [44]). As to monotonicity, coerciveness (inequality (1.5)) and inequalities (2.6), (2.7), (3.1), and (3.2), these hold under suitable additional assumptions.

Now we address some typical situations, restricting ourselves to simple second order operators.

REMARK 7.1. We consider an operator

$$Au = -\sum_{i=1}^n \frac{\partial}{\partial x_i} a\left(\frac{\partial u}{\partial x_i}\right), \tag{7.2}$$

where a: $\mathbf{R}\to\mathbf{R}$ is a continuous monotone increasing function such that

$$c_1\cdot(|\xi|^p-1) \leqslant a(\xi)\cdot\xi \leqslant c_2\cdot(|\xi|^p+1). \tag{7.3}$$

It is not difficult to see that A: $W_0^{1,p}(\Omega)\to W^{-1,p'}(\Omega)$ is a monotone continuous operator. Moreover, (7.3) implies inequalities (1.3) and (1.5). Hence Theorem 5.5 is applicable with $V=W_0^{1,p}(\Omega)$, $H=L^2(\Omega)$ and $V'=W^{-1,p'}(\Omega)$ (we need take into account Remark 5.8 if $V\subset H$).

If

$$\frac{1}{p}-\frac{1}{n}\leqslant\frac{1}{2}, \tag{7.4}$$

then $V \subset H$, and this embedding is compact if we have the strict inequality sign in (7.4) [79]. So in the last case the results of § 4 may be applied. To be able to apply the results of § 2 and § 3 we need additional restrictions. In particular, if the function $a(\xi)$ satisfies the inequality

$$[a(\xi)-a(\eta)]\cdot(\xi-\eta) \geqslant \alpha\,|\,\xi-\eta\,|^q, p \geqslant q \geqslant 2, \tag{7.5}$$

then inequality (3.1) holds with $E = W_0^{1,q}(\Omega)$. But if we have the inequality

$$[a(\xi)-a(\eta)]\cdot(\xi-\eta) \geqslant \alpha\,|\,\xi-\eta\,|^q(1+|\,\xi\,|+|\,\eta\,|)^{p-q}, p < q, \tag{7.6}$$

then (3.2) holds with $E = V = W_0^{1,p}(\Omega)$. For example, the function $a(\xi) = |\,\xi\,|^{p-2}\cdot\xi$ satisfies inequality (7.5) with $q=p$ if $p \geqslant 2$, and inequality (7.6) with $q=2$ if $1<p<2$. The corresponding operator

$$Au = -\sum_{i=1}^{n} \frac{\partial}{\partial x_i} \left\{ \left| \frac{\partial u}{\partial x_i} \right|^{p-2} \cdot \frac{\partial u}{\partial x_i} \right\}$$

is a typical example in the theory of nonlinear elliptic operators.

Now, if (7.4) is not valid we may change the choice of spaces and set $V = W_0^{1,p}(\Omega) \cap L^2(\Omega)$, $H = L^2(\Omega)$ and $V' = W^{-1,p'}(\Omega) + L^2(\Omega)$. It is not true that the operator (7.2) is coercive in the case under the consideration, but the operator defined by

$$Au + cu, c > 0, \tag{7.7}$$

is coercive. Hence we turn out to be in the setting of Remark 1.3 with $V_1 = W_0^{1,p}(\Omega)$ and $V_2 = L^2(\Omega)$. So we may apply the results of § 2 and § 3 (see Theorem 2.9 and Remark 3.13).

For the present we do not concretize the convex function ϕ. Various examples of such functions of interest for applications, and interpretations of corresponding problems, can be found in [21, 44].

EXAMPLE 7.2. We take $H = L^2(\Omega)$ and $V = H^1(\Omega)$, and define the operator A by the bilinear form

$$\alpha(u, v) = \int_\Omega \sum_{i=1}^{n} \frac{\partial u}{\partial x_i} \cdot \frac{\partial v}{\partial x_i} dx, \quad u, v \in H^1(\Omega). \tag{7.8}$$

This operator coincides with $-\Delta$ on $C_0^\infty(\Omega)$. The operator A satisfies inequality

(1.3) with $p=2$, but the coerciveness estimate (1.5) is not valid. Nevertheless, with a special choice of ϕ we can find bounded and a.p. solutions of the corresponding variational inequality. Let

$$K = \{v \in H^1(\Omega) \mid |v(x)| \leqslant h, x \in \partial\Omega\},$$

where $h>0$ is fixed, and let $\phi=\psi_K$ be the indicator function of K, defined by

$$\psi_K(v) = \begin{cases} 0, & v \in K, \\ +\infty, & v \notin K. \end{cases}$$

Inequality (1.2) turns into

$$\int_{t_1}^{t_2} [(v', v-u)+a(u, v-u)-(f, v-u)]dt \geqslant \frac{1}{2} |v-u|^2 \Big|_{t_1}^{t_2}, \tag{7.9}$$

$\forall v \in L^2_{\text{loc}}(\mathbf{R}; V)$, $v' \in L^2_{\text{loc}}(\mathbf{R}; V')$, $v(t) \in K$ for almost all t. The interpretation of the problem is given in [21].

Let $f \in S^2(\mathbf{R}; L^2(\Omega))$. Then for any $u_0 \in \overline{K}^H$ there is a unique solution, u, of the corresponding Cauchy problem with initial data u_0 (see [21]). Since the embedding $H^1(\Omega) \subset L^2(\Omega)$ is compact, then, by Remark 4.10, in order to prove the existence of a solution in the space $\mathrm{CAP}(\mathbf{R}; H) \cap \mathrm{BS}^2(\mathbf{R}; V)$ it suffices to show that $u \in C_b(\mathbf{R}_+; H) \cap \mathrm{BS}^2(\mathbf{R}_+; V)$. Since $u(t) \in K$ for almost all t, we have $\gamma u \in L^\infty(\mathbf{R}_+; L^2(\partial\Omega))$, where γ is the operator of taking the trace of a function on $\partial\Omega$. We now recall that there is a $\beta>0$ such that

$$a(v, v) + \|\gamma v\|^2_{L^2(\partial\Omega)} \geqslant \beta \|v\|^2 \tag{7.10}$$

(see [21], § 7.3 of Ch. 1). Inequality (7.9) for $v=0$ gives rise to the estimate

$$\frac{1}{2} |u|^2 \Big|_{t_1}^{t_2} + \int_{t_1}^{t_2} a(u, u)dt \leqslant \int_{t_1}^{t_2} \|f(t)\|_* \|u(t)\| dt.$$

Taking into account (7.10), we obtain

$$\frac{1}{2} |u|^2 \Big|_{t_1}^{t_2} + \beta \int_{t_1}^{t_2} \|u\|^2 dt \leqslant \int_{t_1}^{t_2} (\|f\|_* \|u(t)\| + \|\gamma u\|^2_{L^2(\partial\Omega)})dt \leqslant$$

$$\leqslant \int_{t_1}^{t_2} (\|f\|_* \|u\| + c\|\gamma u\|_{L^2(\partial\Omega)} \cdot \|u\|)dt \leqslant$$

$$\leqslant \int_{t_1}^{t_2} (\|f\|_* + c\|\gamma u\|_{L^2(\partial\Omega)}) \cdot \|u\| dt.$$

Now the required boundedness of u follows from Lemma 1.1 of Ch. 2 with $\lambda^2 = (1/2)|u|^2$, $\psi = \|u\|$ and $\theta = \|f\|_* + c\|\gamma u\|_{L^2(\partial\Omega)}$. Hence there is an a.p. solution.

EXAMPLE 7.3. Let V, H and A be the same as in Example 7.2. Let

$$\phi(v) = \int_{\partial\Omega} \chi((\gamma v)(x)) d\Gamma, \tag{7.11}$$

where

$$\chi(s) = \begin{cases} 0, & |s| \leqslant h, \\ 1/2k(|s| - h)^2, & |s| > h \end{cases}$$

($k > 0$ and $h > 0$ are fixed). The previous example is the limit case of the present one as $k \to +\infty$. Here also there is a solution in $BS^2(\mathbf{R}; V) \cap CAP(\mathbf{R}; H)$, provided $f \in S^2(\mathbf{R}; L^2(\Omega))$. Indeed,

$$\phi(v) \geqslant \alpha_1 \|\gamma v\|^2_{L^2(\partial\Omega)} + \beta_1,$$

where $\alpha_1 > 0$, $\beta_1 \in \mathbf{R}$, and, by (7.10),

$$\int_{t_1}^{t_2} (a(u, u) + \phi(u)) dt \geqslant \alpha_2 \|u\|^2_{H^1(\Omega)} + \beta_2,$$

where $\alpha_2 > 0$, $\beta_2 \in \mathbf{R}$. Now apply Lemma 1.1 of Ch. 2.

EXAMPLE 7.4. Let V, H and A be the same as in the Example 7.2. For ϕ we take either the indicator function ψ_K of the convex set

$$K = \{v \in H^1(\Omega) \mid (\gamma v)(x) \leqslant h, x \in \partial\Omega\}$$

or the function defined by (7.11), where

$$\chi(s) = \begin{cases} 0, & s \leqslant h, \\ k(s-h)^2/2, & s > h, \end{cases}$$

and $k>0$, $h>0$ are fixed. Let $f \in S^2(\mathbf{R}; L^2(\Omega))$.

Case a): $f(t) \geqslant 0$ for almost all $t \in \mathbf{R}$. By Theorem 6.1 of [21, Ch. 1], a solution of the Cauchy problem with vanishing initial data is nonnegative. Hence $||\gamma u||_{L^2(\partial\Omega)} \leqslant ch$ and, as in Examples 7.2 and 7.3, we see that this solution belongs to $\mathrm{BS}^2(\mathbf{R}_+; V) \cap C_b(\mathbf{R}_+; H)$. As a consequence, there is a solution in $\mathrm{CAP}(\mathbf{R}; H) \cap \mathrm{BS}^2(\mathbf{R}; V)$.

Case b): arbitrary f. Let u_- be a solution on $\mathbf{R}_+$ of the equation

$$\frac{dv}{dt} + Av = f,\ t \in \mathbf{R}_+. \tag{7.12}$$

In other words,

$$\frac{\partial u_-}{\partial t} - \Delta u_- = f,\ t \in \mathbf{R}_+,\ x \in \Omega,$$

$$\frac{\partial u_-}{\partial \nu} = 0,\ t \in \mathbf{R}_+,\ x \in \partial\Omega.$$

Let u be a solution of our inequality with right-hand side f and initial value $u_0(0)$. By Theorem 6.2 of [21, Ch. 1], $u \geqslant u_-$. Now let $f_+ \in S^2(\mathbf{R}; L^2(\Omega))$ be such that $f_+ \geqslant f$ and $f_+ \geqslant 0$. We can find a solution u_+ of our inequality with right-hand side f_+ and initial value $u_+(0) \geqslant 0$, $u_+(0) \geqslant u_-(0)$. By Theorem 6.3 of [21, Ch. 1] we have $u_+ \geqslant u$. The result in case a) implies that $u_+ \in \mathrm{BS}^2(\mathbf{R}_+; V) \cap C_b(\mathbf{R}_+; H)$. Hence our inequality has a solution in $\mathrm{BS}^2(\mathbf{R}_+; V) \cap C_b(\mathbf{R}_+; H)$ and, as a consequence, in $\mathrm{BS}^2(\mathbf{R}; V) \cap \mathrm{CAP}(\mathbf{R}; H)$, provided equation (7.12) has a solution in $\mathrm{BS}^2(\mathbf{R}_+; V) \cap C_b(\mathbf{R}_+; H)$.

We note that if (7.12) has a solution in $\mathrm{BS}^2(\mathbf{R}_+; V) \cap C_b(\mathbf{R}_+; H)$, then any solution of it belongs that space. Indeed, $-\Delta$ with the homogeneous Neumann condition is selfadjoint and nonnegative. Hence all solutions of the corresponding homogeneous equation are bounded on $\mathbf{R}_+$.

In case $\phi = \psi_K$ we have the following formal interpretation of our problem:

$$\frac{\partial u}{\partial t} - \Delta u = f \text{ in } \mathbf{R}\times\Omega,$$

$u-h\leqslant 0$, $\partial u/\partial\nu\leqslant 0$, $(u-h)\cdot\partial u/\partial\nu=0$ on $\partial\Omega$.

Up to sign this problem coincides with the so-called semi-permeable thin wall problem [21, Ch. 1]. Hence we obtain for this problem a sufficient condition for the existence of forced almost periodic oscillations.

7.2. PERTURBATIONS OF VARIATIONAL INEQUALITIES AND PROBLEMS WITH A SHIFTED VARIABLE.

Under the condition of n°. 3.1 and with $q=2\leqslant p$ we consider the variational inequality

$$\int_{t_1}^{t_2}[(v'+Au+Bu-f, v-u)+\phi(v)-\phi(u)]dt \geqslant \frac{1}{2}|v-u|^2\Big|_{t_1}^{t_2} \tag{7.13}$$

for all $t_1\leqslant t_2$ and all test functions v, where the operator B maps $BS^2(\mathbf{R};E)$ into $BS^2(\mathbf{R};E')$ or $C_b(\mathbf{R};H)$ into $BS^2(\mathbf{R};E')$.

PROPOSITION 7.5. *Assume that the conditions of Theorem* 3.1 *with* $q=2\leqslant p$ *are satisfied, and that* B *is a Lipschitz operator with sufficiently small Lipschitz constant* l. *Then for any* $f\in BS^p(\mathbf{R};V')$ *there is a unique solution of inequality* (7.13) *in the space* $BS^p(\mathbf{R};V)\cap C_b(\mathbf{R};H)$.

Proof. By Theorem 3.1,

$$F_f\colon BS^2(\mathbf{R};E')\to BS^2(\mathbf{R};E)$$

is a Lipschitz operator. If B maps $BS^2(\mathbf{R};E)$ into $BS^2(\mathbf{R};E')$ and l is small enough, then the operator

$$F_f\circ B\colon BS^2(\mathbf{R};E)\to BS^2(\mathbf{R};E)$$

is contractive. It is not hard to see that the unique fixed point of this map is a required solution. Similarly, in the second case we consider the operator $F_f\circ B\colon C_b(\mathbf{R};H)\to C_b(\mathbf{R};H)$ and look for its fixed point. □

REMARK 7.6. As in Theorem 3.9 we assume that $A(t)=A_0+A_1(t)$, where $A_1(t)$ satisfies condition $(\mathrm{cap}_{E,2})$ and $f\in S^2(\mathbf{R};E')$. We additionally assume that B maps $S^2(\mathbf{R};E)$ into $S^2(\mathbf{R};E')$ or $CAP(\mathbf{R};H)$ into $S^2(\mathbf{R};E')$. Then the solution constructed in Proposition 7.5 lies in $\mathrm{CAP}(\mathbf{R};H)\cap S^2(\mathbf{R};E)$. Indeed, the fixed point of $F_f\circ B$ may be constructed by successive approximation. Therefore $u\in S^2(\mathbf{R};E)$ or $u\in\mathrm{CAP}(\mathbf{R};H)$. Since by Theorem 3.9 F_f maps $S^2(\mathbf{R};E')$ into $S^2(\mathbf{R};E)\cap\mathrm{CAP}(\mathbf{R};E)$, the required assertion follows from the equality $u=(F_f\circ B)u$.

EXAMPLE 7.7. Let $V=W_0^{1,p}(\Omega)$, $H=L^2(\Omega)$, $E=H$, and

$$Au = -\sum_{i=1}^{n}\frac{\partial}{\partial x_i}\left(\left|\frac{\partial u}{\partial x_i}\right|^{p-2}\cdot\frac{\partial u}{\partial x_i}\right)+\lambda u,\ \lambda>0,\ p\geqslant 2.$$

Define B by the formula

$$(Bv)(t) = \epsilon\cdot v(\omega(t)),$$

where $\epsilon\in\mathbf{R}$ has sufficiently small absolute value, and $\omega:\mathbf{R}\to\mathbf{R}$ is measurable. Then $B: C_b(\mathbf{R};H)\to L^\infty(\mathbf{R};H)\subset \mathrm{BS}^2(\mathbf{R};H)$ is a linear operator whose norm is small enough. Therefore all the conditions of Proposition 7.5 are satisfied. In the case when $\omega(t)=at+s(t)$ and $s\in\mathrm{CAP}(\mathbf{R})$, the conditions of Remark 7.6 are satisfied too.

EXAMPLE 7.8. In the setting of Example 7.7 we assume that $p=2$ and $E=V$. Then $A=-\Delta+\lambda I$. Define B by the formula

$$(Bv)(t,x) = \epsilon_1\cdot\Delta v(\omega_1(t),x)+\epsilon_2\cdot\sum_{i=1}^{n}b_i(x)\frac{\partial v}{\partial x_i}(\omega_2(t),x)+\epsilon_3\cdot v(\omega_3(t),x),$$

where the absolute values of ϵ_i are small enough, and $\omega_i:\mathbf{R}\to\mathbf{R}$ are measurable functions such that B continuously maps $\mathrm{BS}^2(\mathbf{R};V)$ into $\mathrm{BS}^2(\mathbf{R};V')$ $(i=1,2,3)$. Again Proposition 7.5 is applicable. The class of such ω_i's is large enough. For example, this class contains any absolutely continuous function ω for which $\omega'\in L^\infty(\mathbf{R})$. In the case when $\omega_i(t)=a_it+s_i(t)$ and $s_i\in\mathrm{CAP}(\mathbf{R})$ $(i=1,2,3)$, Remark 7.6 is also applicable.

Comments.

§ 1. The general theory of variational inequalities, and applications, is presented in [21, 44, 118]. Here we discuss well-known results. The version of the convergence lemma we have presented here was stated in [67].

§§ 2-4. We follows the paper [67], but present more details. Part of the results has been published also in [57] in a weaker form. One of the first existence theorems for bounded and a.p. solutions dealing with abstract parabolic equations was stated by G. Prouse [145] (see also [100]), provided monotonicity and compactness of the embedding hold. In the same setting V. Zhikov [22] (see also [25]) has obtained a much more general and stronger result. He has also noted a situation in which there is "almost periodicity without compactness". The problem of boundedness and almost periodicity of solutions of variational inequalities has been posed by J.-L. Lions [44] (see Ch. 4, Problem 10.18 and 10.20). He has also obtained partial results in this direction. Subsequently, papers of M. Biroli [107-111] appeared. But in the case of equation (1.6) his results gave rise to weaker results than the theorem of Prouse. At present the existence theorems presented here are the most general results of such kind and completely answer the above-mentioned questions of Lions. In the case of equation (1.6) they give rise to Zhikov's theorem (§ 4) and, on the other hand, to Prouse's theorem, but without any compactness assumption (§ 2 and § 3). We note that up to the appearance of the papers [57, 67] there was no systematic study of regularity of bounded solutions. The single exception is the above-mentioned theorem of Lions in [44, Ch. 4, Theorem 8.4].

§ 5. These results have been obtained in [67] (see also preliminary announcements in [54, 60]). In the case of equation (1.6) V. Zhikov and B. Levitan [25, 41] have also studied Besicovitch a.p. solutions. Their approach is based on completely different ideas. The distinctive feature of our approach is that we introduce a corresponding weak statement of the problem and then use well-developed abstract evolution problem techniques. The approach of [25, 41] uses evolution operators of the corresponding Cauchy problem and, if applicable, gives additional information on solutions. But in the case of Besicovitch a.p. right-hand sides this approach cannot be used directly, because evolution operators are not defined in this case.

§ 6. These results are published for the first time. They are corresponding versions of well-known theorem of Tikhonov [82] (see also [8]) and, in addition, may be considered as generalization of Theorem 9.1 of [44, Ch. 3].

§ 7. Here we consider the simplest examples only. We note that the results of Biroli are applicable in these examples only in a coercive setting. Perturbations of variational inequalities and inequalities with shifted arguments are studied by Biroli [111]. Proposition 7.5 generalizes one of the results of that paper, which itself generalizes other theorems on inequalities with shifted arguments. The results of the present section (except Example 7.1) are published for the first time.

CHAPTER 4.

Bounded and almost periodic solutions of certain evolution equations.

1. Abstract evolution equations.

1.1. BESICOVITCH ALMOST PERIODIC SOLUTIONS.

Consider an evolution equation

$$\frac{du}{dt}+L(t)u+A(t)u = f(t),\ t\in\mathbf{R}, \tag{1.1}$$

with a, generally speaking unbounded linear, operator $L(t)$ and a nonlinear operator $A(t)$. Henceforth this equation will be considered in some weak sense.

We will state assumptions about equation (1.1) and give a weak statement of the problem. Let $\mathscr{E}$ be some locally convex Hausdorff space over $\mathbf{R}$, let $\mathscr{E}'$ be its dual and let V be a reflexive Banach space. We assume that $\mathscr{E}\subset V\subset\mathscr{E}'$ and $\mathscr{E}\subset V'\subset\mathscr{E}'$, all the embeddings being dense and continuous. Suppose that $L(t)$, $t\in\mathbf{R}$, is a family of maximal monotone operators from V to V' with domains $(D(L(t))$. Recall [44] that a closed operator L from V to V' is maximal monotone iff $(Lv, v)\geqslant 0$, $v\in D(L)$, and $(L^*w, w)\geqslant 0$, $w\in D(L^*)$ (L^* is the adjoint in the sense of the theory of unbounded operators). We introduce the following vector spaces:

$$D(\Lambda) = \{u\in \mathrm{CAP}(\mathbf{R}; V) \mid u(\cdot)\in D(L(\cdot)),\ u',\ L(\cdot)u(\cdot)\in \mathrm{CAP}(\mathbf{R}; V')\},$$

$$D(\Lambda^+) = \{u\in \mathrm{CAP}(\mathbf{R}; V) \mid u(\cdot)\in D(L^*(\cdot)),\ u',\ L^*(\cdot)u(\cdot)\in \mathrm{CAP}(\mathbf{R}; V')\}.$$

Here the derivative is interpreted in the weak sense (see Ch. 3, n°. 5.1). On $D(\Lambda)$ and $D(\Lambda^+)$, respectively, we define the operators

$$\Lambda u = \frac{du}{dt}+L(\cdot)u,\ u\in D(\Lambda),$$

$$\Lambda^+ v = -\frac{dv}{dt} + L^*(\cdot)v,\ v \in D(\Lambda^+).$$

They are regarded as unbounded operators from $B^p(\mathbf{R}; V)$ into $B^{p'}(\mathbf{R}; V'), p>1$. We will assume that the following condition is fulfilled:

$(H1)\, D(\Lambda)$ and $D(\Lambda^+)$ are dense in $B^p(\mathbf{R}; V)$.

This assumption includes in particular certain conditions of almost periodicity of $L(t)$, and can be readily verified in a concrete situation. The following statement is obvious.

PROPOSITION 1.1. *The following equality holds:*

$$(\Lambda u, v)_B = (u, \Lambda^+ v)_B,\ u \in D(\Lambda), v \in D(\Lambda^+). \tag{1.2}$$

This implies that the operators Λ and Λ^+ admit closures. We denote by $\mathfrak{L}^*$ a closure of the operator Λ^+ and put $\mathfrak{L} = (\mathfrak{L}^*)^*$. It is obvious that $\mathfrak{L} \supset \Lambda$.

Assume that the following condition is fulfilled:

$(H2)\, D(\Lambda)$ is dense in $D(\mathfrak{L})$ in the graph norm.

This means that, in particular, $\mathfrak{L}$ is a closure of the operator Λ and the operators $\mathfrak{L}$ and $\mathfrak{L}^*$ are mutually adjoint. (This justifies the initial notation $\mathfrak{L}^*$.) By construction, $\mathfrak{L}$ is a maximal (weak) closed extension of the operator Λ and $(H2)$ means that it coincides with the minimal (strong) closed extension.

Now let $A(t): V \to V', t \in \mathbf{R}$, be a family of monotone semicontinuous operators such that

$$\|A(t)v\|_* \leqslant c_1 \|v\|^{p-1} + c_2,\ v \in V, \tag{1.3}$$

$$(A(t)v, v) \geqslant \alpha \|v\|^p + \beta,\ v \in V, \tag{1.4}$$

where $\alpha>0$, $c_1>0$, β, $c_2 \in \mathbf{R}$ do not depend on t. It is also assumed that the following condition is fulfilled:

$(\mathrm{cap}_{V,p})$ The functions $\{(1+\|v\|^{p-1})^{-1} A(t)v\}$ are a.p. uniformly in $v \in V$ and have values in V'.

As in Chapter 3 we may assume, without loss of generality, that $A(\cdot)0=0$ and $\beta=0$.

The following existence theorem holds.

THEOREM 1.2. *Suppose that assumption* $(H1)$, $(H2)$, $(\mathrm{cap}_{V,p})$, *and inequalities* (1.3), (1.4) *are satisfied. Then for any* $f \in B^{p'}(\mathbf{R}; V')$ *there exists a solution*

$u \in B^p(\mathbf{R}; V)$ of the equation

$$\mathfrak{L}u + A(t)u = f(t). \tag{1.5}$$

Proof. By virtue of the results of Ch. 2, n°. 2.2, for $u \in B^p(\mathbf{R}; V)$ the composition $A(t)u(t) \in B^{p'}(\mathbf{R}; V')$, and the operator $\mathbf{A}: B^p(\mathbf{R}; V) \to B^{p'}(\mathbf{R}; V')$ determined by it is monotone and semicontinuous. In addition it satisfies the estimates

$$||\mathbf{A}v||_{B^{p'}} \leqslant C \cdot ||v||_{B^p}^{p-1} + C_1, \; v \in B^p(\mathbf{R}; V),$$

$$(\mathbf{A}v, v)_B \geqslant ||v||_{B^p}^p + \beta, \; v \in B^p(\mathbf{R}; V).$$

Now we will show that

$$(\mathfrak{L}v, v) \geqslant 0, \; v \in D(\mathfrak{L}), \tag{1.6}$$

$$(\mathfrak{L}^* w, w) \geqslant 0, \; w \in D(\mathfrak{L}^*). \tag{1.7}$$

By the definition of $\mathfrak{L}^*$ and condition $(H2)$ it suffices to verify these inequalities for $v \in D(\Lambda)$ and $w \in D(\Lambda^+)$. In this case,

$$(v', v)_B = -(v, v')_B = 0,$$

and by the monotonicity of $L(\cdot)$,

$$(L(\cdot)v, v) \geqslant 0.$$

Thus,

$$(\mathfrak{L}v, v)_B = (\Lambda v, v)_B = (v', v)_B + \mathbf{M}\{(L(\cdot)v, v)\} \geqslant 0.$$

This gives estimate (1.6). Inequality (1.7) is verified similarly.

Estimates (1.6) and (1.7) means that $\mathfrak{L}$ is a maximal operator. Hence to finish the proof of the theorem it suffices to use the assertion: the sum of a maximal monotone operator and a pseudomonotone coercive operator is surjective ([44, Ch. 3, § 1, Theorem 1.1). □

We now consider uniqueness of the solution. Let u_i be the solutions of (1.5) with right-hand sides f_i, $i=1,2$. Then

$$(\mathfrak{L}(u_1-u_2), u_1-u_2)_B+(\mathbf{A}u_1-\mathbf{A}u_2, u_1-u_2)_B = (f_1-f_2, u_1-u_2)_B.$$

By the monotonicity of $\mathfrak{L}$ and $\mathbf{A}$ these give the inequalities

$$0 \leqslant (\mathfrak{L}(u_1-u_2), u_1-u_2)_B \leqslant (f_1-f_2, u_1-u_2)_B, \tag{1.8}$$

$$0 \leqslant (\mathbf{A}u_1-\mathbf{A}u_2, u_1, u_2)_B \leqslant (f_1-f_2, u_1-u_2)_B. \tag{1.9}$$

Hence the following uniqueness theorem holds:

PROPOSITION 1.3. *If in addition to the conditions of Theorem* 1.2 *one of the following conditions is satisfied*:

$$(\mathfrak{L}v, v)_B > 0,\ v\in D(\mathfrak{L}),\ v\neq 0,$$

or

$$(A(t)v-A(t)w, v-w) \geqslant c(v,w) \geqslant 0,\ v,w\in V,$$

where $c(v,w)\neq 0$ *if* $v\neq w$ *and* $c(v,w)$ *does not depend on* t, *then a solution of equation (1.5) is unique.*

REMARK 1.4. The statements about the Hölder property of the inverse operator and continuous dependence of the solution on parameters established in Ch. 3, § 5, remain true in the situation considered, because they are based on estimate (1.9). In the case when L does not depend on t the regularity of solutions in time can be established.

1.2. BOUNDED SOLUTIONS.

Now we proceed to investigate equation (1.1) in more traditional spaces. In order to do this we must restrict somewhat the framework of our considerations: we will assume that $V\subset H\subset V'$, where H is a Hilbert space, $p\geqslant 2$. These restrictions are essential, at any rate for technical reasons. As before, the norm in H is denoted by $|\cdot|$.

As in n°. 1.1, let $L(t)$ be maximal monotone operators from V to V'. We introduce the following vector spaces:

$$D(\Lambda) = \{u\in L^p_{loc}(\mathbf{R}; V) \mid u(\cdot)\in D(L(\cdot)),\ \ L(\cdot)u(\cdot), u'\in L^{p'}_{loc}(\mathbf{R}; V')\},$$

$$D(\Lambda^+) = \{v\in L^p_c(\mathbf{R}; V) \mid v(\cdot)\in D(L^*(\cdot)),\ \ L^*(\cdot)v(\cdot), v'\in L^{p'}_c(\mathbf{R}; V')\},$$

$$D(\Lambda_0, a, b) = \{u\in L^p(a, b; V) \mid u(\cdot)\in D(L(\cdot)),$$

$$L(\cdot)u(\cdot), u'\in L^{p'}(a, b; V'), u(a) = 0\},$$

$$D(\Lambda_0^+, a, b) = \{v\in L^p(a, b; V) \mid v(\cdot)\in D(L^*(\cdot)),$$

$$L^*(\cdot)v(\cdot), v'\in L^{p'}(a, b; V'), v(b) = 0\}.$$

Recall that $L^r_c(\mathbf{R}; E)$, $1\leqslant r\leqslant\infty$, consists of all measurable functions $f{:}\,\mathbf{R}\to E$ such that $\operatorname{supp} f\subset(a, b)$ (modulo a set of measure zero) for some finite interval and $f|_{(a,b)}\in L^r(a, b; E)$. The topology is defined by the neighbourhoods of zero $U(\epsilon, a, b)=\{f \mid \operatorname{supp} f\subset(a, b),\ ||f||_{L^r(a,b)}<\epsilon\}$. Consider the operators

$$\Lambda u = \frac{du}{dt}+L(\cdot)u,\ \ u\in D(\Lambda),$$

$$\Lambda^+ v = -\frac{dv}{dt}+L^*(\cdot)v,\ \ v\in D(\Lambda^+).$$

The operators Λ_0 and Λ_0^+ on the interval (a, b) are defined in a similar way: they have domains $D(\Lambda_0, a, b)$ and $D(\Lambda_0^+, a, b)$ respectively. All these operators are regarded as unbounded on the corresponding spaces. The notation $D(\cdot, a, b)$ for the domain of definition is used every time when it is necessary to indicate the interval on which the differential operator is considered. Note that the operators considered differ from those of n^0. 1.1: they are defined on different spaces.

Assume that the following condition (which is easily verifiable in concrete situations) is fulfilled:

(H3) for any $t_0 \in \mathbf{R}$ the traces of functions from $D(\Lambda)$ at this point are dense in H. For any $a<b$ the spaces $D(\Lambda_0, a, b)$ and $D(\Lambda_0^+, a, b)$ are dense in $L^p(a, b, V)$.

We mention that for $\phi \in C_0^\infty(\mathbf{R})$, $\operatorname{supp}\phi \subset (a, b)$, the inclusion $\phi \cdot D(\Lambda)|_{(a,b)} \subset D(\Lambda_0, a, b)$ holds (similarly for the adjoint operators). This follows from the formula

$$\frac{d}{dt}(\phi u)+L(\cdot)(\phi u) = \phi[\frac{du}{dt}+L(\cdot)u]+\phi' u \tag{1.10}$$

and the fact that $\phi' u \in L^p(a, b; V) \subset L^{p'}(a, b; V')$. (Indeed, $p \geqslant p'$ as $p \geqslant 2$.) Furthermore, under condition $(H3)$ this implies that $D(\Lambda)$ and $D(\Lambda^+)$ are dense in $L^p_{loc}(\mathbf{R}; V)$ and $L^p_c(\mathbf{R}; V)$, respectively. We have also the obvious equalities

$$\int_{\mathbf{R}}(\Lambda u, v)dt = \int_{\mathbf{R}}(u, \Lambda^+ v)dt, \; u \in D(\Lambda), \; v \in D(\Lambda^+), \tag{1.11}$$

$$\int_a^b(\Lambda_0 u, v)dt = \int_a^b(u, \Lambda_0^+ v)dt, \; u \in D(\Lambda_0, a, b), \; v \in D(\Lambda_0^+, a, b).$$

These imply that the operators considered admit closures. Denote by $\mathfrak{L}^*$ and $\mathfrak{L}_0^*$ closures (strong extensions) of the operators Λ^+ and Λ_0^+, respectively, and put $\mathfrak{L}=(\mathfrak{L}^*)^*$, $\mathfrak{L}_0=(\mathfrak{L}_0^*)^*$. The latter are the maximal (weak) closed extensions of Λ and Λ_0. Assume that the following condition is satisfied:

(H4) for any $a<b$ the space $D(\Lambda_0, a, b)$ is dense in $D(\mathfrak{L}_0, a, b)$ in the graph norm.

Formula (1.10) remains true for the operator $\mathfrak{L}$: if $\phi \in C_0^\infty(\mathbf{R})$ and $u \in D(\mathfrak{L})$, then $\phi u \in D(\mathfrak{L})$ and

$$\mathfrak{L}(\phi u) = \phi \mathfrak{L} u+\phi' u \tag{1.12}$$

(the proof is a standard exercise involving adjoint operators). Besides this, $u \in D(\mathfrak{L})$ can be approximated by functions of the form ϕu in the graph topology. It suffices to take $\phi_n \in C_0^\infty(\mathbf{R})$ such that $\phi_n(t)=1$ for $|t| \leqslant n$. Then the desired assertion easily follows from equality (1.12). Therefore $D(\Lambda)$ is also dense in $D(\mathfrak{L})$ in the graph topology. In other words, $\mathfrak{L}$ and $\mathfrak{L}_0$ are closures of Λ and Λ_0, respectively. In addition, the operators $\mathfrak{L}$ and $\mathfrak{L}^*$ are mutually adjoint, as are $\mathfrak{L}_0$ and $\mathfrak{L}_0^*$.

LEMMA 1.5. *The continuous embeddings*

$$D(\mathfrak{L}_0, a, b) \subset C([a, b]; H), \tag{1.13}$$

$$D(\mathfrak{L}) \subset C(\mathbf{R}; H) \tag{1.14}$$

hold, and, moreover, $u(a)=0$ *for* $u \in D(\mathfrak{L}_0, a, b)$. *For* $u \in D(\mathfrak{L})$ *(or* $D(\mathfrak{L}_0, a, b)$ *the inequality*

$$\frac{1}{2} | u(t) |^2 |_{t_1}^{t_2} \leqslant \int_{t_1}^{t_2} (\mathfrak{L}u, u) dt, \ t_1 \leqslant t_2, \tag{1.15}$$

is valid.

Proof. Embedding (1.14) follows from (1.13) and from the statement above formula (1.12).

For $u \in D(\Lambda)$ and $u \in D(\Lambda_0, a, b)$ estimate (1.15) is almost obvious: by the monotonicity of $L(\cdot)$,

$$\int_{t_1}^{t_2} (\mathfrak{L}u, u) dt = \int_{t_1}^{t_2} (u', u) dt + \int_{t_1}^{t_2} (L(t)u(t), u(t)) dt \geqslant \frac{1}{2} | u(t) |^2 |_{t_1}^{t_2}.$$

In the second case $u(a)=0$, and by Hölder's inequality and the previous estimate:

$$|| u ||^2_{C([a,b],H)} \leqslant 2 || \mathfrak{L}u ||_{L^{p'}(a,b;V')} || u ||_{L^p(a,b;V)} \leqslant$$

$$\leqslant || u ||^2_{L^p(a,b;V)} + || \mathfrak{L}u ||^2_{L^{p'}(a,b;V')},$$

that is, on the space $D(\Lambda_0, a, b)$ the graph norm of $\mathfrak{L}_0$ is stronger that the $C([a, b]; H)$-norm. Since $D(\Lambda_0, a, b)$ is dense in $D(\mathfrak{L}_0, a, b)$, this implies the continuity of (1.13).

As was stated above, estimate (1.15) holds on $D(\Lambda_0, a, b)$, and therefore on $\phi \cdot D(\Lambda)$ for all $\phi \in C_0^\infty(\mathbf{R})$. The general case is obtained from these by passing to the limit with the aid of $(H4)$ and the continuity of the embeddings (1.13), (1.14), which was established above. □

REMARK 1.6. An assertion similar to Lemma 1.5 is valid for the adjoint operators.

Concerning the operators $A(t): V \to V'$ we assume that estimates (1.3), (1.4) are satisfied and that the following condition is fulfilled:

(c) *for any* $v \in V$ *and any bounded set* $U \subset V$ *the family of functions* $\{(A(\cdot)u, v) \mid u \in U\}$ *is equicontinuous on every compact interval.*

Equation (1.1) will be understood in a generalized sense, as an equation

$$\mathfrak{L}u + A(t)u = f. \tag{1.16}$$

As follows from Lemma 1.5, for any two solutions u_i of this equation with right-hand sides f_i, $i = 1, 2$, respectively, the following estimate holds:

$$\frac{1}{2} | u_1 - u_2 |^2 |_{t_1}^{t_2} + \int_{t_1}^{t_2} (Au_1 - Au_2, u_1 - u_2) dt \leqslant \int_{t_1}^{t_2} (f_1 - f_2, u_1 - u_2) dt. \tag{1.17}$$

It is similar to the estimate given by Lemma 1.5 (Ch. 3).

Here, as in Ch. 3, we need a statement about passage to the limit in equation (1.16).

LEMMA 1.7. *Let* u_n *be a solution of equation (1.16) with right-hand side* f_n. *Suppose that* $\lim f_n = f$ *strongly in* $L^{p'}_{\mathrm{loc}}(\mathbf{R}; V')$, $\lim u_n = u$ *weakly in* $L^p_{\mathrm{loc}}(\mathbf{R}; V)$ *and strongly in* $C(\mathbf{R}; H)$. *Then* $\mathfrak{L}u + A(t)u = f$.

Proof. By results of Ch. 2, n°. 2.1, for any $a < b$ the sequence $\{Au_n\}$ is bounded in $L^{p'}(a, b; V')$. Hence, passing over to a subsequence if necessary, we may assume that $\lim A(t)u_n(t) = g(t)$ weakly in $L^{p'}_{\mathrm{loc}}(\mathbf{R}; V')$. From the equation

$$\mathfrak{L}u_n + A(t)u_n = f_n$$

It follows that the sequence $\{\mathfrak{L}u_n\}$ is bounded in any $L^{p'}(a, b; V')$. Thus we may assume that $\{\mathfrak{L}u_n\}$ converges weakly in $L^{p}_{\mathrm{loc}}(\mathbf{R}; V')$. Since the notions of weak and strong closedness of a linear operator coincide*, this implies that $u \in D(\mathfrak{L})$ and $\lim \mathfrak{L}u_n = \mathfrak{L}u$ weakly in $L^{p'}_{\mathrm{loc}}(\mathbf{R}; V')$.

Hence

$$\mathfrak{L}u + g = f,$$

and it remains to show that

$$g(t) = A(t)u(t). \tag{1.18}$$

* For convex subsets (in particular for graphs of linear operators) in locally convex spaces, being weakly closed is equivalent to being closed in the ordinary topology [88].

The relation

$$\mathfrak{L}(u_n-u)+A(t)u_n-g(t) = f_n-f$$

and inequality (1.15) give

$$\frac{1}{2}|u_n-u|^2|_a^b+\int_a^b(Au_n-g, u_n-u)dt \leqslant \int_a^b(f_n-f, u_n-u)dt$$

for any $a<b$. Therefore, by the convergence of the sequences $\{u_n\}$ and $\{f_n\}$,

$$\limsup\int_a^b(Au_n-g, u_n-u)dt \leqslant \limsup\int_a^b(Au_n, u_n-u)dt \leqslant 0.$$

Hence

$$\limsup\int_a^b(Au_n, u_n)dt \leqslant \lim\int_a^b(Au_n, u)dt = \int_a^b(g, u)dt.$$

Now note that composition with $A(t)$ is a monotone semicontinuous operator from $L^p(a, b; V)$ into $L^{p'}(a, b; V')$, and consequently is an operator of the type (M) (see [44], Ch. 2, Proposition 2.5). Hence (1.18) follows from (1.19). □

REMARK 1.8. Lemma 1.7 can be generalized to the case when equation (1.16) includes a nonlinear term $A_n(t)\rightarrow A(t)$ in the metric $d_{V,p}$ (see Ch. 2, n°. 2.2) uniformly in t, i.e. if

$$\sup_{t\in\mathbf{R},v\in V}\frac{||A_n(t)v-A(t)v||_*}{1+||v||^{p-1}}\rightarrow 0$$

(compare with Ch. 3, Corollary 1.7).

THEOREM 1.9. *Suppose that conditions* $(H3)$,$(H4)$, (c), *inequalities* (1.3), (1.4), *and the estimate*

$$(A(t)v-A(t)w, v-w) \geqslant \alpha|v-w|^p, \quad v, w\in V, p\geqslant 2, \tag{1.20}$$

are satisfied. Then equation (1.16) *has a unique solution* $u\in BS^p(\mathbf{R}; V)\cap C_b(\mathbf{R}; H)$ *for any* $f\in BS^{p'}(\mathbf{R}; V')$.

Proof. We will follows the proof of Theorem 2.5 (Ch. 3). Set $f_n = \chi_n f$, where χ_n is the characteristic function of the half-line $[-n, +\infty)$, and consider a solution of the Cauchy problem

$$\mathfrak{L}u_n + Au_n = f_n,\ t \in \mathbf{R},\ u_n(t) = 0,\ t \leqslant -n. \tag{1.21}$$

Results in [104] imply that such a solution exists. Estimates (1.4), (1.15) give the inequality

$$\frac{1}{2}|u_n|^2\Big|_{t_1}^{t_2} + \alpha\int_{t_1}^{t_2} \| u_n \|^p dt \leqslant \int_{t_1}^{t_2} (f_n, u_n)dt,\ t_1 \leqslant t_2.$$

From this and Lemma 1.1 (Ch. 2) we find that $u_n \in \mathrm{BS}^p(\mathbf{R}; V) \cap C_b(\mathbf{R}; H)$, and the sequence $\{u_n\}$ is bounded in this space.

In view of Lemma 1.7 it suffices to establish the existence of the strong limit $\lim u_n = u$ in $C(\mathbf{R}; H)$ in order to prove existence of the solution. (Since V is reflexive, $\lim u_n = u$ weakly in $L^p_{\mathrm{loc}}(\mathbf{R}; V)$.) Equality (1.21) and estimates (1.17), (1.20) give the inequality

$$\frac{1}{2}|u_n - u_m|^2\Big|_{t_1}^{t_2} + \alpha\int_{t_1}^{t_2} | u_n - u_m |^p dt \leqslant \int_{t_1}^{t_2} (f_n - f_m, u_n - u_m)dt.$$

Since $f_n(t) = f_m(t) = f(t)$ for $t \geqslant t_{n,m} = -\min(m, n)$, by using Lemma 1.3 (Ch. 2) as in the proof of Theorem 2.5 (Ch. 3) we obtain that $\{u_n\}$ is a Cauchy sequence in $C(\mathbf{R}; H)$. Its limit obviously lies in $\mathrm{BS}^p(\mathbf{R}; V) \cap C_b(\mathbf{R}; H)$ and so existence of the solution is proved.

Uniqueness is proved in the same way as in Theorem 2.4 (Ch. 3). □

REMARK 1.10. The norms of the solution just constructed can be estimated in the following way

$$\| u \|_{C_b(\mathbf{R}; H)} \leqslant C_1(\| f \|_{S^{p'}}),$$

$$\| u \|_{S^p} \leqslant C_2(\| f \|_{S^{p'}}),$$

where $C_i(x)$, $x \geqslant 0$, $i = 1, 2$, are continuous nondecreasing functions depending only on c_1, c_2, α, and β. □

1.3. BOHR AND STEPANOV ALMOST PERIODIC SOLUTIONS.

With equation (1.16) we associate the operator $F: f \to u$, where u is a bounded solution.

LEMMA 1.11. *Under the conditions of Theorem* 1.9 *the operator* $F: \mathrm{BS}^{p'}(\mathbf{R}; V') \to C(\mathbf{R}; H)$ *is continuous. If* (1.20) *is replaced by the stronger inequality*

$$(A(t)v - A(t)w, v-w) \geqslant \alpha \| v-w \|^p, \ v, w \in V,$$

then the operators $F: \mathrm{BS}^{p'}(\mathbf{R}; V) \to C_b(\mathbf{R}; H)$ *and* $F: \mathrm{BS}^{p'}(\mathbf{R}; V') \to \mathrm{BS}^{p}(\mathbf{R}; V)$ *are locally Hölder with exponents* $(p-1)^{-1}$ *and* $2/p(p-1)$, *respectively.*

Proof. The proof of Theorem 3.1 (Ch. 3) is based on the four stages: construction of approximate solutions, estimation of the difference between two different solutions (in our case estimate (1.17)), the lemma about the limit passage, and Lemma 1.1 (Ch. 2).

Hence the proof can be immediately carried over to the situation considered. □

In order to study the almost periodicity of the solutions constructed in Theorem 1.9 we must impose severe restrictions on the operators $L(t)$:

(H5) $L(t) = L_0 + L_1(t)$, *where* L_0 *is a closed operator from* V *to* V' *and* $L_1(t)$ *is an a.p. function taking values in the space* $L(V, V')$ *of all bounded linear operators.*

The operator function $L_1(t)$ can be extended to a continuous function on $\mathbf{R}_B$ with values in $L(V, V')$, for which we retain the previous notation. Thus we may define on operator $\mathfrak{L}_s$ as a closure of the operator

$$\frac{d}{dt} + L_0 + L_1(\cdot + s), \ s \in \mathbf{R}_B.$$

Its domain does not depend on s, and for all operators assumptions $(H3)$, $(H4)$ hold if these hold for the operator $\mathfrak{L}$.

THEOREM 1.12. *Suppose that assumptions* $(H3)$-$(H5)$, $(\mathrm{cap}_{V,p})$, *and inequalities* (1.3), (1.4), *and* (1.20) *hold for* $p \geqslant 2$. *Then for* $f \in S^{p'}(\mathbf{R}; V')$ *the unique bounded solution* u *of equation (1.16) lies in* $\mathrm{CAP}(\mathbf{R}; H)$. *If* (1.22) *holds instead of* (1.20), *then* $u \in S^p(\mathbf{R}; V) \cap \mathrm{CAP}(\mathbf{R}; H)$.

Proof. The operator-valued function $A(t)$ is extended to $\mathbf{R}_B$ as a function continuous in the $d_{V,p}$-metric. The function $f(\cdot + s)$ is a.p. in s and hence can also be extended to a function continuous on $\mathbf{R}_B$, with values in $\mathrm{BS}^{p'}(\mathbf{R}; V')$. Now consider the family of equations

$$\mathfrak{L}_s u_s + A(t+s)u_s = f(t+s),\ s \in \mathbf{R}_B. \tag{1.23}$$

Theorem 1.9 is applicable to any such equation. Therefore, for any $s \in \mathbf{R}_B$ there is a unique solution $u_s \in \mathrm{BS}^p(\mathbf{R}; V) \cap C_b(\mathbf{R}; H)$ $(u_0 = u$ is a solution of (1.16)). By uniqueness,

$$u_s(\cdot) = u(\cdot + s),\ s \in \mathbf{R} \subset \mathbf{R}_B. \tag{1.24}$$

It is easy to see that

$$u_s = F_{s_0}[f(\cdot+s) + (L_1(\cdot+s_0) - L_1(\cdot+s))u_s +$$

$$+ A(\cdot+s_0)u_s - A(\cdot+s)u_s], \tag{1.25}$$

where F_s is the inverse of the operator corresponding to equation (1.23). By Remark 1.10 the set $\{u_s\}$ is bounded in the space $\mathrm{BS}^p(\mathbf{R}; V)$. Hence, by $(\mathrm{cap}_{V,p})$, $\lim[A(\cdot+s_0)u_s - A(\cdot+s)u_s] = 0$ in $\mathrm{BS}^{p'}(\mathbf{R}; V')$, and by $(H5)$,

$$\lim_{s \to s_0} [L_1(\cdot+s_0) - L_1(\cdot+s)]u_s = 0 \text{ in } \mathrm{BS}^p(\mathbf{R}; V')$$

and, consequently, in $\mathrm{BS}^{p'}(\mathbf{R}; V')$ (since $p \geqslant 2$). By Lemma 1.11 this and representation (1.25) imply that u_s is continuous in $s \in \mathbf{R}_B$ as an element of $C(\mathbf{R}; H)$. We put $u(s) = u_s(0)$, $s \in \mathbf{R}_B$, to obtain the continuous extension of $u(t)$ to $\mathbf{R}_B$. Consequently, $u \in \mathrm{CAP}(\mathbf{R}; H)$.

In order to prove the second assertion of the theorem we use the second part of Lemma 1.11. Then $u_s \in \mathrm{BS}^p(\mathbf{R}; V)$ depends continuously on $s \in \mathbf{R}_B$. This fact together with (1.24) gives $u \in S^p(\mathbf{R}; V)$. □

REMARK 1.13. There is an obvious parallel between the results of Chapter 3, § 3 and the results presented above. In particular, for the solutions of equation (1.16) the theorem about continuous dependence on parameters and regularity in t may be proved (the latter in the case when $L(t) = L_0 + L_1(t)$, where L_1 is a bounded operator which sufficiently regularly depends on t). Also, an intermediate space E may be introduced and inequality (1.22) may be replaced by inequalities (3.1), (3.2). But from the point of view of the applications which will be considered in § 2, such a generalization is not so interesting. The matter is that a situation with estimate (3.2) appears, as a rule, in the case $1 < p < 2$; a case which is not considered here (except in n°. 1.1) for other reasons.

REMARK 1.14. Assume that the conditions of Theorem 1.12 except estimate (1.20) are fulfilled. Replacing $A(t)$ by $A(t)+\lambda I, \lambda>0$, in equation (1.16) we may apply Theorem 1.12 and obtain a solution of the modified equation $u_\lambda \in CAP(\mathbf{R}; H)$ which, in addition, belongs to $B^p(\mathbf{R}; V)$ (see Ch. 1, Theorem 5.14). The arguments of the kind used to prove Proposition 5.4 (Ch. 3) allows us to pass to the limit as $\lambda \to 0$ and to obtain a solution $u \in B^p(\mathbf{R}; V) \bigcap B^\infty(\mathbf{R}; H)$ of equation (1.1) in the sense of n°. 1.1. Besides, in this situation we can use the arguments developed in [25], n°. 9.9 (see also [41], § 4, Ch. IX) and construct a function $u_0 \in B^p(\mathbf{R}; V) \bigcap B^2(\mathbf{R}; H)$ such that for almost all $s \in \mathbf{R}_B$ the function $u_0(s+t), t \in \mathbf{R}$, is well-defined and is a bounded solution of (1.23). Also, it can be shown that u_0 is a solution in the sense of n°. 1.1.

REMARK 1.15 We have used only the monotonicity methods above. If $D(\mathfrak{L})$ is compactly embedded in some $L^r_{\text{loc}}(\mathbf{R}; H)$, then the techniques developed in Chapter 3, § 4 may be adopted to get rid of condition (1.29) in Theorems 1.9 and 1.12 (but possibly uniqueness may be lost). Unfortunately, in applications of interest to us such compactness conditions are, as a rule, not satisfied.

2. Applications.

2.1. SYMMETRIC HYPERBOLIC SYSTEMS WITH A MONOTONE NONLINEARITY.

Let $\Omega \subset \mathbf{R}^n$ be a bounded domain with boundary of class C^1. We consider a system of equations

$$\frac{\partial u}{\partial t} + \sum_{i=1}^{n} L_i(t, x)\frac{\partial u}{\partial x_i} + M(t, x)u + a(t, x, u) = f(t, x), \tag{2.1}$$

where $(t, x) \in \mathbf{R} \times \Omega$, L_i, $i=1, \ldots, n$, are real symmetric $(N \times N)$- matrices of class $C^2(\mathbf{R} \times \overline{\Omega})$, and M is a real matrix of class $C(\mathbf{R} \times \Omega)$. We assume that the operator

$$L(t) = \sum_{i=1}^{n} L_i(t, x)\frac{\partial}{\partial x_i} + M(t, x) \tag{2.2}$$

is formally positive, that is,

$$M(t, x) + M^*(t, x) - \sum_{i=1}^{n} \frac{\partial L_i}{\partial x_i}(t, x) \geqslant 0, \ (t, x) \in \mathbf{R} \times \overline{\Omega}. \tag{2.3}$$

Let $B(t,x)$ be an $(N\times q)$-matrix $(q\leqslant N)$ of class $C^1(\mathbf{R}\times\partial\Omega)$ and of constant rank q. We look for a solution u of (2.1) satisfying the boundary condition

$$B(t,x)u(t,x) = 0,\ (t,x)\in\mathbf{R}\times\Omega. \tag{2.4}$$

Let $\nu(x)=(\nu_1(x),\ldots,\nu_n(x))$ be the outward normal unit vector to $\partial\Omega$ and put

$$L^{(\nu)}(t,x) = \sum_{i=1}^{n} L_i(t,x)\nu_i(x),\ (t,x)\in\mathbf{R}\times\partial\Omega.$$

It is assumed that the matrix $L^{(\nu)}(t,x)$ is invertible for any $(t,x)\in\mathbf{R}\times\partial\Omega$. Let

$$\mathfrak{B}(t,x) = \ker B(t,x) = \{\xi\in\mathbf{R}^N \mid B(t,x)\xi = 0\}$$

be the space of boundary conditions. We assume that the boundary conditions are positive, that is,

$$\xi\cdot L^{(\nu)}(t,\xi)\xi \geqslant 0,\ \forall\xi\in\mathfrak{B}(t,x),\ (t,x)\in\mathbf{R}\times\partial\Omega,$$

and satisfy the maximality condition: for any subspace $E\subset\mathbf{R}^N$ strictly including $\mathfrak{B}(t,x)$ there is a vector $\eta\in E$ such that $\eta\cdot L^{(\nu)}(t,x)\eta<0$ (the dot denotes the scalar product in $\mathbf{R}^n$). Under these conditions the differential operator (2.2) together with the boundary conditions (2.4) generates a maximal monotone operator $L(t)$ that acts from $L^p(\Omega)$ to $L^{p'}(\Omega), p\geqslant 2$ [104] (see also [1, 128, 129] for the classical case $p=2$), Note that L acts on the spaces of vector-valued functions $L^r(\Omega,\mathbf{R}^N)=L^r(\Omega)^N$. To denote this spaces we use the same notation as in the scalar case. The adjoint operator $L^*(t)$ is generated by the operator

$$-\sum_{i=1}^{n}\frac{\partial}{\partial x_i}L_i(t,x)+M^*(t,x)$$

with boundary conditions

$$C(t,x)v(t,x) = 0,\ (t,x)\in\mathbf{R}\times\partial\Omega.$$

Here $C(t,x)$ is an $(N\times(N-q))$-matrix of constant rank $N-q$, and the space of conjugate boundary conditions,

$$\mathfrak{C}(t,x) = \ker C(t,x),$$

is orthogonal to $L^{(\nu)}(t,x)\mathfrak{B}(t,x)$.

Next we assume that $a:\mathbf{R}\times\Omega\times\mathbf{R}^N\to\mathbf{R}^n$ is a continuous mapping such that

$$|a(t,x,\xi)| \leqslant c_1|\xi|^{p-1}+c_2,\ (t,x,\xi)\in\mathbf{R}\times\Omega\times\mathbf{R}^N,$$

$$a(t,x,\xi)\cdot\xi \geqslant \alpha|\xi|^p+\beta,\ (t,x,\xi)\in\mathbf{R}\times\Omega\times\mathbf{R}^N,$$

$$[a(t,x,\xi_1)-a(t,x,\xi_2)]\cdot(\xi_1-\xi_2) \geqslant 0,\ (t,x)\in\mathbf{R}\times\Omega,\ \xi_1,\xi_2\in\mathbf{R}^N,$$

where the constants $c_1>0$, $\alpha>0$, and $c_2,\beta\in\mathbf{R}$ do not depend on (t,x,ξ).

A typical example is $a(t,x,\xi)=|\xi|^{p-2}\xi$.

Equation (2.1) is understood in the sense of § 1, that is, as equation (1.5) or (1.16). In case (1.16) this is the usual interpretation of the problem in the sense of distributions. In case (1.5) equation (2.1) is satisfied, for example, in the space $\mathfrak{D}'(\mathbf{R}_B;W^{-1,p'}(\Omega))$.

The main result concerning problem (2.1), (2.4) is

THEOREM.2.1. *Under the above assumptions the following statements are true:*

a) if

$$[a(t,x,\xi_1)-a(t,x,\xi_2)]\cdot(\xi_1-\xi_2) \geqslant \alpha|\xi_1-\xi_2|^p,$$

$$(t,x)\in\mathbf{R}\times\Omega,\ \xi_1,\xi_2\in\mathbf{R}^N, \tag{2.5}$$

and $f\in BS^{p'}(\mathbf{R};L^{p'}(\Omega))$, *then problem* (2.1), (2.4) *has a unique solution* $u\in BS^p(\mathbf{R};L^p(\Omega))\cap C_b(\mathbf{R};L^2(\Omega))$;

b) if $L_i\in CAP^2(\mathbf{R};C^2(\bar\Omega))$, $i=1,\dots,n$, $M\in CAP(\mathbf{R};C(\bar\Omega))$, $B\in CAP^1(\mathbf{R};C^1(\partial\Omega))$, *and* $a(t,x,\xi)/(1+|\xi|^{p-1})$ *is a.p. in t uniformly with respect to* $(x,\xi)\in\Omega\times\mathbf{R}^N$, *then for* $f\in B^{p'}(\mathbf{R};L^{p'}(\Omega))$ *problem* (2.1), (2.4) *has a solution* $u\in B^p(\mathbf{R};L^p(\Omega))$;

c) if in addition to the assumption of b), $L_i(t,x)=L_i(x)$, $i=1,\dots,n$, $B(t,x)=B(x)$ *and* (2.5) *holds, then for* $f\in S^{p'}(\mathbf{R};L^{p'}(\Omega))$ *problem* (2.1), (2.4) *has a unique solution* $u\in S^p(\mathbf{R};L^p(\Omega))\cap CAP(\mathbf{R};L^2(\Omega))$.

Proof. To prove the results we verify the assumptions of Theorem 1.2, 1.9 and 1.12 with $V=L^p(\Omega)$, $H=L^2(\Omega)$.

Validity of conditions (c) and $(\mathrm{cap}_{V,p})$ in the situation considered follows from results of Ch. 2, § 2.

Verification of conditions $(H1)$ and $(H3)$ is fairly simple. In the case $(H1)$, $D(\Lambda)$ and $D(\Lambda^+)$ obviously contain the subspace $\mathrm{CAP}^1(\mathbf{R}; W_0^{1,p}(\Omega))$, which is dense in $B^p(\mathbf{R}; L^p(\Omega))$. In the case $(H3)$, $D(\Lambda_0, a, b)$ (respectively, $D(\Lambda_0^+, a, b)$) contains the subspace $\{v \in C^1([a, b]; W_0^{1,p}(\Omega)) \mid v(a)=0$ (respectively, $v(b)=0)\}$, which is dense in $L^p(a, b, L^p(\Omega))$. Note that $L(\cdot): W_0^{1,p}(\Omega) \to L^p(\Omega)$ is a bounded operator.

In the case $p=2$ condition $(H4)$ is satisfied, due to the classical result that weak and strong solutions coincide [1, 128, 129]. In case $p>2\,(H4)$ is established in [104, Theorem IV. 2] with the aid of a suitable generalization of classical techniques.

Condition $(H2)$ will be verified later.

LEMMA 2.2. *For any open cover of* $\mathbf{R}_B^n$ *there exists a subordinate partition of unity consisting of functions from* $\mathrm{CAP}^\infty(\mathbf{R}^n)$.

Proof. The group $\mathbf{R}_B^n$ can be represented as the projective limit of a system of tori,

$$\mathbf{R}_B^n = \lim\operatorname{proj} \mathbf{T}^{N_\lambda} \tag{2.6}$$

(see. Ch. 1, § 3). By the standard construction of a partition of unity [51] it is sufficient to prove that for any point $a \in \mathbf{R}_B^n$ and any neighbourhood Y of it there is a function $\phi_{a,Y} \in \mathrm{CAP}^\infty(\mathbf{R}^n)$ such that $\phi_{a,Y} \geqslant 0$, $\phi_{a,Y}(a)>0$ and $\operatorname{supp}\phi_{a,Y} \subset Y$. Now observe that the inverse images under canonical epimorphisms $p_\lambda: \mathbf{R}_B^n \to \mathbf{T}^{N_\lambda}$ of neighbourhoods of the points $p_\lambda(a)$ in $\mathbf{T}^{N_\lambda}$ constitute a fundamental system of neighbourhoods of the point a in $\mathbf{R}_B^n$. Choose a neighbourhood U in $\mathbf{T}^{N_\lambda}$ such that $p_\lambda^{-1}(U) \subset Y$. Then there is a function $\psi \in C^\infty(\mathbf{T}^{N_\lambda})$ such that $\psi \geqslant 0$, $\psi(p_\lambda(a))>0$ and $\operatorname{supp}\psi \subset U$. Now it remains to put $\phi_{a,Y} = p_\lambda^* \psi = \psi \circ p_\lambda$. □

LEMMA 2.3. *There is a δ-shaped net* $\{\theta_Y\} \subset \mathrm{CAP}^\infty(\mathbf{R}^n)$ *enumerated by a system* $\mathcal{Y}$ *of neighbourhoods of zero in* $\mathbf{R}_B^n$ *such that* $\theta_Y \geqslant 0$ *and* $\operatorname{supp}\theta_Y \subset Y$ *for all* $Y \subset \mathcal{Y}$.

Proof. In order to construct the $\{\theta_Y\}$ it is sufficient to normalize the functions $\phi_{0,Y}$ in the proof of the previous lemma,

$$\theta_Y(a) = \phi_{0,Y}(x)/\mathbf{M}\{\phi_{0,Y}\}. \square$$

Soon we will need some results (similar to well-known results in [1, 128, 129]) about smoothing operators in the case of functions which are a.p. with respect to part of the variables. Consider the integral operator

$$(\Phi_\epsilon u)(x) = \int \phi_\epsilon(x, y, x-y) u(y) dy, \quad x \in \mathbf{R}^n. \tag{2.7}$$

It is assumed that $\phi_\epsilon(x, y, \eta)$ is continuous in (x, y, η) and almost periodic in x_1, y_1

uniformly with respect to $x_2, \ldots, x_n, y_2, \ldots, y_n, \eta$, that $\phi_\epsilon(x,y,\eta)=0$ if $|\eta| \geqslant \epsilon$, and that

$$\int |\phi_\epsilon(x,y,x-y)|\,dx \leqslant C, \ \int |\phi_\epsilon(x,y,x-y)|\,dy \leqslant C,$$

where C does not depend on ϵ, x, y. Moreover, let

$$c_\epsilon(x) = \int \phi_\epsilon(x,y,x-y)dy \to c$$

uniformly in $x \in \mathbf{R}^n$. Then the following assertion holds:

LEMMA 2.4. *The operators* Φ_ϵ *are uniformly bounded in the space* $B^r(\mathbf{R}; L^r(\mathbf{R}^{n-1})) = L^r(\mathbf{R}_B \times \mathbf{R}^{n-1})$, *and* $\lim_{\epsilon\to 0} || \Phi_\epsilon u - cu || = 0$.

Proof. It is not difficult to see that Φ_ϵ is a bounded operator on $C_b(\mathbf{R}; L^r(\mathbf{R}^{n-1}))$. Moreover it acts on $CAP(\mathbf{R}; L^r(\mathbf{R}^{n-1}))$. Indeed, Φ_ϵ can be represented as an integral operator whose kernel $\psi_\epsilon(x_1, y_1, x_1-y_1)$ is an operator-valued function acting on $L^r(\mathbf{R}^{n-1})$. Then

$$|| (\Phi_\epsilon u)(x_1,\cdot) - (\Phi_\epsilon u)(x_1+h,\cdot) ||_{L^r} \leqslant$$

$$\leqslant \int || \psi_\epsilon(x_1,y_1,x_1-y_1) - \psi_\epsilon(x_1+h, y_1+h, x_1-y_1) || \cdot || u(y_1,\cdot) ||_{L^r} dy_1 +$$

$$+ \int || \psi_\epsilon(x_1+h, y_1+h, x_1-y_1) || \cdot || u(y_1,\cdot) - u(y_1+h,\cdot) ||_{L^r} dy_1.$$

The operator-valued function $\psi_\epsilon(x_1, y_1, z)$ is a.p. in x_1, y_1 uniformly with respect to z in the norm topology and equals zero when $|z| \geqslant \epsilon$. Therefore, if h is a common δ-almost period of the functions $\psi_\epsilon(x_1, y_1, \cdot)$ and $u(y_1, \cdot)$, then the previous estimate gives

$$|| (\Phi_\epsilon u)(x_1,\cdot) - (\Phi_\epsilon u)(x_1+h,\cdot) ||_{L^2} \leqslant \delta \int_{|x_1-y_1| \leqslant \epsilon} || u(y_1,\cdot) ||_{L^r} dy_1 +$$

$$+ C_1 \int_{|x_1-y_1| \leqslant \epsilon} || u(y_1,\cdot) - u(y_1+h,\cdot) ||_{L^r} dy_1 \leqslant C_2 \cdot \delta,$$

that is, h is a $C_2 \cdot \delta$-almost period of the function $(\Phi_\epsilon u)(x_1, \cdot)$ in the $L^r(\mathbf{R}^{n-1})$-norm.

Next, using Hölder's inequality and Fubini's theorem we have, for $u \in \mathrm{CAP}(\mathbf{R}; L^r(\mathbf{R}^{n-1}))$,

$$\frac{1}{2}\int_{-T}^{T} \| (\Phi_\epsilon u)(x_1, \cdot) \|_{L^r}^r dx_1 \leqslant$$

$$\leqslant \frac{1}{2T}\int_{-T}^{T} [\int \| \psi_\epsilon(x_1, y_1, x_1 - y_1) \| \cdot \| u(y_1, \cdot) \|_{L^r} dy_1]^r dx_1 \leqslant$$

$$\leqslant \frac{1}{2T}\int_{-T}^{T} [\int \| \psi_\epsilon(x_1, y_1, x_1 - y_1) \| dy_1]^{r/r'} \times$$

$$\times [\int_{-T-\epsilon}^{T+\epsilon} \| \psi_\epsilon(x_1, y_1, x_1 - y_1) \| \, \| u(y_1, \cdot) \|_{L^r}^r dy_1] dx_1 \leqslant$$

$$\leqslant C^{1+r/r'} \cdot \frac{1}{2T} \int_{-T-\epsilon}^{T+\epsilon} [\int_{-T}^{T} \| \psi_\epsilon(x_1, y_1, x_1 - y_1) \| dx_1] \| u(y_1, \cdot) \|_{L^r}^r dy_1 \leqslant$$

$$\leqslant C^{r/r'} \left[\frac{T+\epsilon}{T} \right] \frac{1}{2(T+\epsilon)} \int_{-T-\epsilon}^{T+\epsilon} \| u(y_1, \cdot) \|_{L^r}^r dy_1.$$

Passing to the limit as $T \to \infty$ we obtain the inequality

$$\| \Phi_\epsilon u \|_{B^r} \leqslant C \| u \|_{B^r}.$$

In particular, Φ_ϵ may be continuously extended to a family of uniformly bounded operators in $B^r(\mathbf{R}; L^r(\mathbf{R}^{n-1}))$.

Now due to the uniform boundedness of Φ_ϵ just proved it is sufficient to establish the last assertion of the lemma for $u \in \mathrm{CAP}(\mathbf{R}; L^r(\mathbf{R}^{n-1}))$. Then $\lim \| \Phi_\epsilon u - cu \| = \lim \| \Phi_\epsilon u - c_\epsilon u \|$ (it is obvious that $c_\epsilon(x)$ is a.p. in x) and

$$(\Phi_\epsilon u - c_3 u)(x_1, \cdot) = \int_{|x_1 - y_1| \leqslant \epsilon} \psi_\epsilon(x_1, y_1, x_1 - y_1)[u(y_1, \cdot) - u(x_1, \cdot)] dy_1.$$

This gives

$$||(\Phi_\epsilon u - c_\epsilon u)(x_1,\cdot)||_{L^r} \leq C \sup_{|x_1-y_1|\leq\epsilon} ||u(x_1,\cdot)-u(y_1,\cdot)||_{L^r}.$$

Hence $||(\Phi_\epsilon u - c_\epsilon u)(x_1,\cdot)||_{L^r}\to 0$ uniformly in x_1 and consequently in $B^r(\mathbf{R}; L^r(\mathbf{R}^{n-1}))$. □

REMARK 2.5. If the kernel is continuously differentiable and all its derivatives are a.p. in x_1, y_1 uniformly with respect to $x_2, \ldots, x_n, y_2, \ldots, y_n, \eta$, then it is easy to see that $\partial/\partial x_i(\Phi_\epsilon u)\in B^r(\mathbf{R}; L^r(\mathbf{R}^{n-1}))$, $i=1,\ldots,n$. In particular $\Phi_\epsilon u \in B^r(\mathbf{R}; W^{1,r}(\mathbf{R}^{n-1}))$.

End of the proof of Theorem 2.1. To verify condition $(H2)$ it is necessary to show that any element $u\in D(\mathfrak{L})$ can be approximated in the graph norm by elements from $D(\Lambda)$. First of all we note that existence of a smooth partition of unity in $\overline{\Omega}$ (under the condition $p\geq 2$) permits us to assume that $\operatorname{supp} u \subset \mathbf{R}_B\times U$, where U is a sufficiently small open subset in $\mathbf{R}^n$. We have two cases.

The case $U\cap\partial\Omega=\varnothing$ is rather simple. Let Φ_ϵ be the standard Friedrichs averages with respect to the variables x [1, 128, 129]. They are uniformly bounded in $L^r(\mathbf{R}^n)$, and consequently in $B^r(\mathbf{R}; L^r(\mathbf{R}^n))$. The functions of the form $v=\psi\cdot v_1$, where $\psi\in C_0^\infty(\Omega)$, $v_1\in \mathrm{CAP}^1(\mathbf{R}; W_0^{1,p}(\Omega))$ lies in $D(\Lambda^+)$, are dense in $D(\mathfrak{L}^*)$ in the graph norm and $\Phi_\epsilon v\in\mathrm{CAP}^1(\mathbf{R}; W_0^{1,p}(\Omega))$ for small ϵ (depending on $\operatorname{supp}\psi$). Furthermore, $\operatorname{supp}\Phi_\epsilon u\subset\mathbf{R}_B\times U$ (also for small ϵ). For such v we have

$$(\Phi_\epsilon\mathfrak{L}u, v)_B = (\mathfrak{L}u, \Phi_\epsilon v)_B = (u, \Lambda^+\Phi_\epsilon v)_B =$$

$$= (\Phi_\epsilon u, \Lambda^+ v)_B + (u, [L^*(\cdot), \Phi_\epsilon]v)_B, \tag{2.8}$$

where $[A,B]=AB-BA$ is the commutator. Without loss of generality we may assume that in these equalities $\operatorname{supp} v\subset\mathbf{R}_B\times U_1$ and $\operatorname{supp}\Phi_\epsilon u\subset\mathbf{R}_B\times U_1$, where $\overline{U}\subset U_1$, $U_1\cap\partial\Omega=\varnothing$. The operators $[L^*(t),\Phi_\epsilon]$ are uniformly bounded in $L^2(\mathbf{R}^n)$ with respect to t and ϵ. Hence $[L^*(\cdot),\Phi_\epsilon]$ is uniformly bounded in $B^2(\mathbf{R}; L^2(\mathbf{R}^n))$ with respect to ϵ. Therefore $(\Phi_\epsilon u, \Lambda^+ v)_B$ and $(u, [L^*(\cdot), \Phi_\epsilon]v)_B$ are linear functionals that are continuous (with respect to v) in the norm of $B^p(\mathbf{R}; L^p(\mathbf{R}^n))$. By (2.8) the same is true for $(\Phi_\epsilon u, \Lambda^+ v)_B$. Thus, $\phi_\epsilon u\in D(\mathfrak{L})$ and

$$\Phi_\epsilon\mathfrak{L} = \Phi_\epsilon - [L(\cdot), \Phi_\epsilon]. \tag{2.9}$$

Now we have $[L(\cdot),\Phi_\epsilon]u\to 0$ in $B^2(\mathbf{R}; L^2(\mathbf{R}^n))\subset B^{p'}(\mathbf{R}; L^{p'}(\mathbf{R}; L^{p'}(\mathbf{R}^n))$, $\Phi 0u\to u$ in

$B^p(\mathbf{R}; L^p(\mathbf{R}^n))$ and $\Phi_\epsilon \mathfrak{L} u \to \mathfrak{L} u$ in $B^{p'}(\mathbf{R}; L^{p'}(\mathbf{R}^n))$. Therefore (2.9) implies that the $\Phi_\epsilon u$ approximate u in the graph norm. Since Φ_ϵ are smoothing operators in the Sobolev scale, $\Phi_\epsilon \in B^p(\mathbf{R}; W_0^{1,p}(\Omega))$ and $\Phi_\epsilon \mathfrak{L} u \in B^{p'}(\mathbf{R}; L^{p'}(\Omega))$. This together with (2.8) implies that $(\Phi_\epsilon u)' \in B^{p'}(\mathbf{R}; L^{p'}(\Omega))$. Hence we may assume that $u \in B^p(\mathbf{R}; W_0^{1,p}(\Omega))$, $u' \in B^{p'}(\mathbf{R}; L^{p'}(\Omega))$.

Now we apply the Bochner-Fejér operators L_γ with respect to the variable t. Then $L_\gamma u \in \mathrm{CAP}^\infty(\mathbf{R}; W_0^{1,p}(\Omega)) \subset D(\Lambda)$ and $\lim L_\gamma u = u$ in $B^p(\mathbf{R}; W_0^{1,p}(\Omega)) \subset B^p(\mathbf{R}; L^p(\Omega))$ and $\lim_\gamma (L_\gamma u)' = u'$ in $B^{p'}(\mathbf{R}; L^{p'}(\Omega))$. Consequently, the $L_\gamma u$ approximate u in the graph norm.

The case $U \cap \partial\Omega$. In view of the existence of a CAP^∞-partition of unity (Lemma 2.2) we may assume that $\operatorname{supp} u \subset W \times U$, where W is an arbitrary subset of $\mathbf{R}_B$. Changing variables, if necessary, we may reduce the situation to the case when $\Omega = \mathbf{R}^n_+ = \{x = (x', x_n) \mid x_n > 0\}$. Since the rank of $B(t, x)$ is constant, it follows (possibly after shrinking W and U) that we may introduce new unknowns so that in $W \times U$ the boundary conditions have the form

$$u^1(t, x', 0) = \cdots = u^q(t, x', 0) = 0, \tag{2.10}$$

the adjoint boundary conditions are

$$v^{q+1}(t, x', 0) = \cdots = v^N(t, x', 0) = 0, \tag{2.11}$$

and the operator $\mathfrak{L}$ can be written as

$$\frac{\partial}{\partial t} + H_n \frac{\partial}{\partial x_n} + \sum_{j=1}^{n-1} H_j \frac{\partial}{\partial x_j} + K. \tag{2.12}$$

Without loss of generality we may assume that the matrix $H(t, x', 0)$ is diagonal and invertible and that $K \equiv 0$. Furthermore, we may assume that H_n and H_n^{-1} are bounded in $W \times U$.

Now we introduce new function spaces. Let $W^{1,p}(\mathbf{R}_B \times \mathbf{R}^n)$, $1 < p < \infty$, be the completion of $\mathrm{CAP}^1(\mathbf{R}; W^{1,p}(\mathbf{R}^n))$ with respect to the norm

$$\| f \|^p = \mathbf{M}\{\| f(\cdot) \|^p_{W^{1,p}}\} + \mathbf{M}\{\| \frac{\partial f}{\partial t}(\cdot) \|^p_{L^p}\}.$$

Define $W^{-1,p'}(\mathbf{R}_B \times \mathbf{R}^n)$, $1/p + 1/p' = 1$, as the dual of $W^{1,p}(\mathbf{R}_B \times \mathbf{R}^n)$, where the duality is given by the formula

$$(u, v)_B = \mathbf{M}_t\{\int u(t, x) v(t, x) dx\},$$

which must be understood in the following sense. If $E = W^{1,p}(\mathbf{R}^n) \cap L^2(\mathbf{R}^n)$, then $\mathfrak{D}(\mathbf{R}_B, E)$ is continuously and densely embedded in $W^{1,p}(\mathbf{R}_B \times \mathbf{R}^n)$. therefore $W^{-1,p'}(\mathbf{R}_B \times \mathbf{R}^n)$ is continuously and densely embedded in the space $\mathfrak{D}'(\mathbf{R}_B; E')$, and $(\cdot, \cdot)_B$ is a restriction of the duality described in Ch. 3, n°. 5.1. First order differential operators act in these spaces exactly in the same manner as in the usual Sobolev spaces (compare with Ch. 1, n°. 6.2).

From the representation (2.12) it follows that

$$\frac{\partial u}{\partial x_n} \in L^{p'}(\mathbf{R}_{+,x_n}; W^{-1,p'}(\mathbf{R}_B \times \mathbf{R}^{n-1})).$$

Thus $u \in C(\overline{\mathbf{R}}_{+,x_n}; W^{-1,p'}(\mathbf{R}_B \times \mathbf{R}^{n-1}))$, and the following "Green's formula" holds:

$$\int_0^\infty (H_n \frac{\partial u}{\partial x_n}, v)_B dx_n = -\int_0^\infty (u, \frac{\partial}{\partial x_n} H_n v)_B dx_n - (u(\cdot, x_n), H_n(\cdot, x_n) v(\cdot, x_n))_B \,|_{x_n=0}$$

for $v \in \mathrm{CAP}^1(\mathbf{R}; C^1(\overline{\mathbf{R}}^n_+))$, $\operatorname{supp} v \subset W \times U$. If $v \in D(\Lambda^+)$, then

$$(\mathfrak{L}u, v) = (u, \Lambda^+ v) - (u(\cdot, x_n), H_n(\cdot, x_n) v(\cdot, x_n))_B \,|_{x_n=0}.$$

By the definition of $\mathfrak{L}$, the second term on the right-hand side vanishes. Since the matrix $H_n(t, x', 0)$ is diagonal and invertible and since $v(t, x', 0)$ satisfies conditions (2.11), $u(t, x', 0)$ satisfies (2.10) (in the sense of the space $W^{-1,p'}(\mathbf{R}_B \times \mathbf{R}^{n-1})$).

Now we choose $\phi(t, x') \in C_0^\infty(\mathbf{R}_t \times \mathbf{R}^{n-1})$ in such a way that $\phi(t, x') \geqslant 0$, $\int \phi(t, x') dt dx' = 1$ and put $\phi_\epsilon(t, x') = \epsilon^{-n} \phi(t/\epsilon, x'/\epsilon)$. Lemma 2.4 may be applied to the operators $\Phi_\epsilon = \phi_\epsilon \star$ (convolution in (t, x')). Moreover, for sufficiently small ϵ the action of Φ_ϵ does not take $\operatorname{supp} u$ out of the set $W \times U$. Indeed, for convolution with respect to x' this is well-known while for convolution with respect to t this follows from the fact that the action of the group $\mathbf{R}$ on $\mathbf{R}_B$ by translation is continuous and the fact that $\operatorname{supp} \phi_\epsilon$ is small in $\mathbf{R}_t \times \mathbf{R}^{n-1}$. For such ϵ the action of Φ_ϵ obviously preserves boundary condition (2.10), (2.11).

Next we have

$$H_n^{-1} \cdot \mathfrak{L} \Phi_\epsilon u = \Phi_\epsilon H_n^{-1} \mathfrak{L} u + [\mathfrak{L}', \Phi_\epsilon] u, \tag{2.13}$$

where

$$\mathfrak{L}' = H_n^{-1} \left[\frac{\partial}{\partial t} + \sum_{j=1}^{n-1} H_j \frac{\partial}{\partial x_j} \right].$$

By Lemma 2.4, $\Phi_\epsilon u \to 0$ in $B^p(\mathbf{R}; L^p(\Omega))$ and $\Phi_\epsilon \mathfrak{L} u \to \mathfrak{L} u$ in $B^{p'}(\mathbf{R}; L^{p'}(\Omega))$. An immediate verification shows that $[\mathfrak{L}', \Phi_\epsilon]$ is an integral operator of the form (2.7) with kernel

$$\Psi_\epsilon(\hat{x}, \hat{y}, \hat{x} - \hat{y}) = \sum_{j=0}^{n-1} \frac{\partial}{\partial y_i} \{(G_j(\hat{y}) - G_j(\hat{x}))\phi_\epsilon(\hat{x} - \hat{y})\},$$

where $\hat{x} = (x_0, x')$, $\hat{y} = (y_0, y')$, $G_0 = H_n^{-1}$, $G_j = H_n^{-1} H_j$, $j = 1, \ldots, n-1$, and instead of t the notation x_0 is used. This kernel depends on the parameter x_n. Since ψ_ϵ is a sum of derivatives with respect to y_j of functions which are, as functions of $\hat{y}$, of compact support, we have

$$c_\epsilon(\hat{x}) = \int \psi_\epsilon(\hat{x}, \hat{y}, \hat{x} - \hat{y}) d\hat{y} = 0.$$

The application of Lemma 2.4 shows that $\lim[\mathfrak{L}', \Phi_\epsilon]u = 0$ in $B^p(\mathbf{R}; L^p(\Omega)) \subset B^{p'}(\mathbf{R}; L^{p'}(\Omega))$. Since H_n and H_n^{-1} are bounded on $W \times U$, this together with (2.13) gives that $\lim \Phi_\epsilon u = u$ in the graph norm. By virtue to Remark 2.5,

$$\Phi_\epsilon u \in L^p(\mathbf{R}_{+,x_n}; W^{1,p}(\mathbf{R}_B \times \mathbf{R}^{n-1})), \quad \frac{\partial \Phi_\epsilon u}{\partial x_n} \in B^{p'}(\mathbf{R}; L^{p'}(\mathbf{R}^n_+)).$$

Hence we may assume that the functions u and $\partial u / \partial x_n$ themselves lie in the spaces just specified:

$$u \in L^p(\mathbf{R}_{+,x_n}; W^{1,p}(\mathbf{R}_B \times \mathbf{R}^{n-1})), \quad \frac{\partial \mu}{\partial x_n} \in B^{p'}(\mathbf{R}; L^{p'}(\mathbf{R}^n_+)). \tag{2.14}$$

Now, taking into account (2.14), we apply to u the smoothing operator K_ϵ with respect to the variable x_n with diagonal matrix kernel

$$k_\epsilon(x_n) = \operatorname{diag}(k_\epsilon^{(1)}(x_n), \ldots, k_\epsilon^{(N)}(x_n)),$$

where $k_\epsilon^{(j)} \in C_0^\infty(\mathbf{R})$ are nonnegative functions with unit integrals and

$$\operatorname{supp} k_\epsilon^{(j)} \subset \begin{cases} \{y \mid \epsilon < y < 2\epsilon\}, & j = 1, \ldots, q, \\ \{y \mid -2\epsilon < y < -\epsilon\}, & j = q+1, \ldots, n \end{cases}$$

(see [1]). For ϵ small enough the action of this operator preserves boundary conditions and does not take $\operatorname{supp} u$ out of $W \times U$. Moreover, $\lim K_\epsilon u = u$ and $\lim \frac{\partial}{\partial x_n} K_\epsilon u = \frac{\partial u}{\partial x_n}$ in the spaces (2.14), and hence also in the graph norm. Moreover, it is obvious that

$$K_\epsilon u, \frac{\partial}{\partial x_n} K_\epsilon u \in L^p(\mathbf{R}_+; W^{1,p}(\mathbf{R}_B \times \mathbf{R}^{n-1})),$$

and hence $K_\epsilon u \in W^{1,p}(\mathbf{R}_B \times \mathbf{R}^n_+)$. Consequently it may be assumed that u belongs to this space.

In order to approximate u by elements from $D(\Lambda)$ we preform another smoothing with respect to variable t. Let $\{\theta_Y\}$ be the kernels constructed in Lemma 2.3. Let $u_\psi = \theta_Y *_B u$ (convolution in $\mathbf{R}_B$). Then since $\{\theta_Y\}$ is smooth and δ-shaped it follows that $u_Y \in \mathrm{CAP}^\infty(\mathbf{R}; W^{1,p}(\mathbf{R}^n_+))$ and $\lim u_Y = u$ in $W^{1,p}(\mathbf{R}_B \times \mathbf{R}^n_+)$. Furthermore, $\operatorname{supp} u_Y \subset W \times U$ and u_Y satisfies the boundary conditions (2.10) for sufficiently small neighbourhoods of zero Y. Consequently, $u_Y \in D(\Lambda)$ and $\lim u_Y = u$ in $D(\mathfrak{L})$. □

REMARK 2.6. In the above proof we have used double regularization with respect to t: first by means of usual convolution, then by means of convolution in $\mathbf{R}_B$. An attempt to eliminate the first regularization and only perform the second, which at first glance seems to be simpler, leads to serious difficulties in the estimation of the commutators of the corresponding smoothing operators and can be achieved only under serious restrictions on the coefficients of $L(t, x)$, such as quasi-periodicity in t. The problems lies in the fact that locally in $\mathbf{R}_B$ (like in a torus with densely embedded line) there are next to the coordinate t "transversal" variables hardly controllable under commutation. The same situation is described in Ch. 5, n°. 2.1. □

2.2. NONLINEAR SCHRÖDINGER-TYPE EQUATION.

Consider the problem

$$\frac{\partial u}{\partial t} - i\Delta u + a(u) = f(t, x), \ (t, x) \in \mathbf{R} \times \Omega, \tag{2.15}$$

$$u|_{\partial\Omega} = 0, \tag{2.16}$$

where $\Omega \subset \mathbf{R}^n$ is a bounded region with smooth boundary (as in n°. 2.1), and f and

u are complex-valued functions. Here $a{:}\mathbf{C}\rightarrow\mathbf{C}$ is a continuous function and $(p>1)$,

$$(a(z)-a(w))\cdot(\bar{z}-\bar{w}) \geqslant 0,$$

$$|\, a(z) \,| \leqslant c_1 |\, z \,|^{p-1} + c_2,$$

$$a(z)\cdot\bar{z} \geqslant \alpha |\, z \,|^p + \beta,$$

for $z, w \in \mathbf{C}$, where $c_1>0$, $\alpha<0$, $c_2, \beta \in \mathbf{R}$. A typical example is $a(z)=|\, z \,|^{p-2}z$. Let us apply to problem (2.15), (2.16) the results of § 1 with $H=L^2(\Omega)$, $V=L^p(\Omega)$. These spaces of complex-valued functions will be regarded as real Banach spaces.

THEOREM 2.7. *a) For* $f \in B^{p'}(\mathbf{R}; L^{p'}(\Omega))$ *problem* (2.15), (2.16) *has a solution* $u \in B^p(\mathbf{R}; L^p(\Omega))$ *(which is unique if* $a(z)$ *is a strictly monotone function).*

b) Let $p \geqslant 2$ *and*

$$(a(z)-a(w))\cdot(\bar{z}-\bar{w}) \geqslant \alpha |\, z-w \,|^p, \quad z, w \in \mathbf{C}.$$

Then for $f \in \mathrm{BS}^{p'}(\mathbf{R}; L^{p'}(\Omega))$ *problem* (2.15), (2.16) *has a unique solution* $u \in \mathrm{BS}^p(\mathbf{R}; (\Omega)) \cap C_b(\mathbf{R}; L^2(\Omega))$. *This solution belongs to* $S^p(\mathbf{R}; L^p(\Omega) \cap \mathrm{CAP}(\mathbf{R}; L^2(\Omega))$ *if* $f \in S^{p'}(\mathbf{R}; L^{p'}(\Omega))$.

Proof. The operator $-i\Delta$ with Dirichlet type boundary conditions generates a skewsymmetric operator L from $L^p(\Omega)$ to $L^{p'}(\Omega)$. Hence L is maximal monotone. The verification of conditions $(H1)$ and $(H3)$ is immediate and is thus left to the reader. The fact that $(H4)$ holds was established in [44, Ch. 3, n°. 2.5] and [104]. We show that $(H2)$ holds (under the conditions of part a)).

Let $u \in D(\mathfrak{L})$. The Bochner-Fejér operators obviously commute with $\partial/\partial u - i\Delta$. Thus, $u=\lim L_\gamma u$ in the graph norm. Hence it may be assumed that $u \in \mathrm{CAP}^\infty(\mathbf{R}; L^p(\Omega))$, $u' - i\Delta u \in \mathrm{CAP}^\infty(\mathbf{R}; L^{p'}(\Omega))$.

Now let $p \geqslant 2$. Then $u' \in \mathrm{CAP}^\infty(\mathbf{R}; L^{p'}(\Omega))$, and consequently $\Delta u \in \mathrm{CAP}^\infty(\mathbf{R}; L^p(\Omega))$. However, under the boundary conditions (2.16) the operator Δ is invertible and $\Delta^{-1} L^r(\Omega) \rightarrow W^{2,r}(\Omega)$, $1<r<\infty$, is a bounded operator [97]. Thus (if $r=p'$ in the previous statement)

$$u \in \mathrm{CAP}^\infty(\mathbf{R}; W^{2/p'}(\Omega)) \cap \mathrm{CAP}^\infty(\mathbf{R}; L^p(\Omega)), \tag{2.17}$$

and consequently $u \in D(\Lambda)$.

If $1<p<2$, then $p<p'$ and $\Delta u \in CAP^{\infty}(\mathbf{R}; L^p(\Omega))$. We have $u \in CAP^{\infty}(\mathbf{R}; W^{2,p}(\Omega))$ because Δ is invertible in $L^p(\Omega)$). By Sobolev's embedding theorem [79] $u \in CAP^{\infty}(\mathbf{R}; L^{p_1}(\Omega))$, where $p_1^{-1} = p^{-1} - 2n^{-1}$ (p_1 may be taken arbitrarily if the right-hand side of this equality is nonpositive). Hence $u' \in CAP^{\infty}(\mathbf{R}; L^{p_1}(\Omega))$. In case $p_1 \geqslant p'$ we have $u' \in CAP^{\infty}(\mathbf{R}; L^{p'}(\Omega))$, from which (2.17) follows. In case $p_1 < p'$, by replacing p by p_1 in the previous argument we obtain $u \in CAP^{\infty}(\mathbf{R}; W^{2,p_1}(\Omega)) \subset CAP^{\infty}(\mathbf{R}; L^{p_2}(\Omega))$, where $p_2^{-1} = p_1^{-1} - 2n^{-1} = (p_2$ is arbitrary if $p_2^{-1} - 4n^{-1} \leqslant 0)$. This procedure may be continued if $p_2 < p'$. In this way we obtain a sequence of indices p_k, where $p_k^{-1} = p^{-1} - 2kn^{-1}$ and p_k is arbitrary if $p^{-1} - 2nk^{-1} \leqslant 0$. It is clear that $p_k \geqslant p$ for sufficiently large k, hence (2.17) holds. □

REMARK 2.8. Of course we may consider a nonlinear term $a(t, x, u)$ which depends in a suitable way on (t, x). The operator $-\Delta$ may be replaced by an arbitrary nonnegative selfadjoint second order elliptic operator.

2.3. PARABOLIC PROBLEMS.

Choosing $V = L^p(\Omega)$, $H = L^2(\Omega)$ and $L = -\Delta$ with domain of definition given by the Neumann conditions, we can apply the results of § 1 to the problem

$$\frac{\partial u}{\partial t} - \Delta u + |u|^{p-2} \cdot u = f(t, x),\ (t, x) \in \mathbf{R} \times \Omega, \tag{2.18}$$

$$\frac{\partial u}{\partial \nu}\Big|_{\partial\Omega} = 0. \tag{2.19}$$

This leads to the following Theorem.

THEOREM 2.9. *If* $f \in B^{p'}(\mathbf{R}; L^{p'}(\Omega))$, *then problem* (2.18), (2.19) *has a unique solution* $u \in B^p(\mathbf{R}; L^p(\Omega))$. *If* $p \geqslant 2$ *and* $f \in BS^{p'}(\mathbf{R}; L^{p'}(\Omega))$, *then there exists a unique solution* $u \in BS^p(\mathbf{R}; L^p(\Omega)) \cap C_b(\mathbf{R}; L^2(\Omega))$. *This solution lies in* $S^p(\mathbf{R}; L^p(\Omega)) \cap CAP(\mathbf{R}; L^2(\Omega))$ *if* $f \in S^{p'}(\mathbf{R}; L^{p'}(\Omega))$.

We note that results of Ch. 3 (and papers [22, 25, 41]), which are connected with the representation of the problem (2.18), (2.19) in the form of equation $u' + A(u) = f$, where $A(u) = -\Delta u + |u|^{p-2} \cdot u$, are not applicable. Indeed, the form

$$\int_{\Omega} \sum_{i=1}^{n} \left[\frac{\partial u}{\partial x_i}\right]^2 dx$$

has nontrivial kernel in $H^1(\Omega)$. Therefore the corresponding operator $A:H^1(\Omega)\bigcap L^p(\Omega)\to[H^1(\Omega)]'+L^{p'}(\Omega)$ (the choice of spaces is natural in this case) will be noncoercive. Certainly, the results of the § 1 may be applied to equation (2.18) with Dirichlet boundary conditions, but in this case the results of Ch. 3 can be applied too.

We also mention the paper [72], in which a.p. solutions of certain nonlinear parabolic equations are constructed on the basis of compactness considerations without using monotonicity. The techniques of this paper are closely related to those of [144] for the two-dimensional Navier-Stokes equation.

3. Additional results.

3.1. COUPLED EQUATIONS.

Consider two reflexive Banach spaces V_1, V_2 where $V_i\subset H\subset V_i'$, $i=1,2$ (as usual, the embeddings are dense and continuous), and H is a Hilbert space. Suppose we are given families of monotone semicontinuous operators $A_i(t):V_i\to V_i'$, $t\in\mathbf{R}$, $i=1,2$, such that

$$||A_i(t)v||_{*i}\leqslant c_1||v||_i^{p_i-1}+c_2,\ v\in V_i, \tag{3.1}$$

$$(A_i(t)v,v)\geqslant\alpha||v||_i^{p_i}+\beta,\ v\in V_i, \tag{3.2}$$

where $p_1\geqslant 2$, $c_1,\alpha>0$, $c_2,\beta\in\mathbf{R}$ do not depend on $t\in\mathbf{R}$. It is assumed that the condition (cap_{V,p_i}) is fulfilled for A_i. Consider the following system of equations:

$$\frac{du_1}{dt}+A_1(t)u_1+u_2=f_1, \tag{3.3}$$

$$-\frac{du_2}{dt}+A_2(t)u_2-u_1=f_2. \tag{3.4}$$

A natural two-point problem for this system was studied in [44, Ch. 2, § 7]. For a.p. solutions there is the following theorem (the derivatives are understood in the sense of Ch. 3, § 5.1):

THEOREM 3.1. *Assume that* $f_i \in B^{p_i'}(\mathbf{R}; V_i')$, $i=1,2$. *Then the system* (3.3), (3.4) *has a solution* $u=(u_1, u_2)$ *such that* $u_i \in B^{p_i}(\mathbf{R}; V_i)$, $i=1,2$.

Proof. Let $\mathcal{V}=B^{p_1}(\mathbf{R}; V_1)\times B^{p_2}(\mathbf{R}; V_2)$; then $\mathcal{V}'=B^{p_1'}(\mathbf{R}; V_1')\times B^{p_2'}(\mathbf{R}; V_2')$. Consider the operator $\mathbf{A}:\mathcal{V}\to\mathcal{V}'$ defined by the relation

$$(\mathbf{A}u)(t) = (A_1(t)u_1(t),\ A_2(t)u_2(t)).$$

By the results of Ch. 2, n°. 2.2, this is a bounded semicontinuous monotone and coercive operator. A simple verification shows that the operator $\mathbf{A}+\mathbf{B}$, where $(\mathbf{B}u)(t)=(u_2(t), -u_1(t))$, has the same properties. The operator $(\mathbf{L}u)(t)=(u_1'(t), -u_2'(t))$ with domain

$$D(\mathbf{L}) = \{u=(u_1, u_2)\in\mathcal{V} \mid u_i' \in B^{p_i'}(\mathbf{R}; V_i'), i=1,2\}$$

is a skewsymmetric operator from $\mathcal{V}$ to $\mathcal{V}'$ which is, moreover, maximal monotone (unbounded). Since the system (3.3), (3.4) is equivalent to the equation

$$\mathbf{L}u+(\mathbf{A}+\mathbf{B})u = f,$$

it remains to apply a standard result about surjectivity of a sum [44, Ch. 3, Theorem 1.1]. □

REMARK 3.2. Usually there is a result on uniqueness of the solution. For this, we only need to strengthen the property of monotonicity, requiring that A satisfies the estimates

$$(A_i(\cdot)v - A_i(\cdot)w, v-w) \geqslant c(v,w) \geqslant 0,$$

where $c(v{\cdot}w) \neq 0$ if $v \neq w$ and does not depend on t.

The question about the existence of bounded and Stepanov (or Bohr) a.p. solutions is essential more complicated, and there is a corresponding result only under string additional assumptions. In particular, only the linear case has been considered.

Let $V_1=V_2=V$ be a Hilbert space and $A(t)\in L(V, V')$ a bounded family of continuous operators such that

$$(A(t)v, v) \geqslant \alpha \,||\, v \,||^2,\ \alpha > 0,\ v\in V \tag{3.5}$$

(this correspondens to the case $p_1=p_2=2$ above). Consider the system of equations:

$$\frac{du_1}{dt}+A(t)u_1+u_2 = f_1, \tag{3.6}$$

$$-\frac{du_2}{dt}+A^*(t)u_2-u_1 = f_2. \tag{3.7}$$

Under these assumptions there is the following results:

THEOREM 3.3. *For* $f_i \in BS^2(\mathbf{R}; V_i')$ *there is a unique solution* $u=(u_1, u_2)$ *of problem* (3.6), (3.7) *for which* $u_i \in BS^2(\mathbf{R}; V) \cap C_b(\mathbf{R}; H)$. *If, moreover,* $A(t)$ *is a.p. with respect to t in the operator norm and* $f_i \in S^2(\mathbf{R}; V')$, *then* $u_i \in S^2(\mathbf{R}; V) \cap CAP(\mathbf{R}; H)$.

Proof. 1) *Uniqueness.* Since the problem is linear, it is sufficient to show that the corresponding homogeneous problem ($f_i \equiv 0$) has only the trivial solution. Multiplying (3.6) by u_1, (3.7) by u_2 and adding them we obtain

$$\frac{1}{2}(|u_1|^2-|u_2|^2)'+(Au_1, u_1)+(A^*u_2, u_2) = 0.$$

By integrating and using (3.5) we find

$$\frac{1}{2}(|u_1|^2-|u_2|^2)|_{t_1}^{t_2}+\alpha\int_{t_1}^{t_2}(|u_1|^2+|u_2|^2)dt \leqslant 0. \tag{3.8}$$

This together with the boundedness of u_i implies the convergence of the integral

$$\int_{-\infty}^{\infty}(|u_1|^2+|u_2|^2)dt.$$

In particular, there is a sequence $\tau_{\pm k} \to \pm\infty$ such that

$$\lim_{k\to\infty}(|u_1(\tau_{\pm k})|^2+|u_2(\tau_{\pm k})|^2) = 0.$$

If we put $t_1=t_{-k}, t_2=\tau_k$ in (3.8) and pass to the limit, then

$$\int_{-\infty}^{\infty}(|u_1|^2+|u_2|^2)dt = 0,$$

and consequently $u \equiv 0$.

2) *Existence*. Without loss of generality we may assume that $f_1 \equiv 0$. Indeed, consider a solution $v \in BS^2(\mathbf{R}; V) \bigcap C_b(\mathbf{R}; H)$ of the equation

$$\frac{dv}{dt} + A(t)v = f_1$$

(its existence and uniqueness follows from results of Ch. 3, for example). We look for solutions of the form $u=(u_1, u_2)$, where $u_1 = v+w$. Then the functions w and u_2 must satisfy the system (3.6), (3.7) with $f_1 \equiv 0$. Of course, f_2 changes but it remains in $BS^2(\mathbf{R}; V')$.

Let us consider a solution $u_n = (u_{1n}, u_{2n})$ of the system (3.6), (3.7) on the interval $[-n, n]$ satisfying the boundary conditions

$$u_{1n}(-n) = 0, \; u_{2n}(n) = 0. \tag{3.9}$$

Such a solution exists and is unique. Furthermore, $u_{in} \in L^2(-n, n, V) \bigcap C([-n, n]; H)$ ([45, Ch. 3, n°. 4.)]). We want to obtain estimates in suitable norms for these solutions, uniform with respect to n. For a function $f \in L^r(a, b; E)$ the S^r-norm is defined by

$$||f||_{S^r} = \sup_{\substack{0 \leqslant t_2 - t_1 \leqslant 1 \\ a \leqslant t_1 \leqslant t_2 \leqslant b}} \left\{ \int_{t_1}^{t_2} ||f(\tau)||_E^r d\tau \right\}^{1/r}.$$

This norm is analogous to the norm of BS^r on the line or on the half-line. Of course, this norm is equivalent to the usual L^r-norm, but the corresponding constants involved depend on the length of the interval (a, b). Temporarily we will drop the index n in the notation of the solution constructed.

We use the results of [45] and Ch. 3, § 4. It is not difficult to see that the problem (3.6), (3.7), (3.9) corresponds to an optimal control problem with $D_1 = D_2 = I$ (in the notation of [45]). Hence it admits so-called decoupling. This means that there exist operators $P(t) \in L(H)$, $t \in [-n, n]$, and a function $r(t)$ such that

$$u_2(t) = P(t)u_1(t) + r(t). \tag{3.10}$$

We state some properties of P and r needed in the sequel.

The operator function $P(t)$ is continuous with respect to t in the weak operator topology and is uniformly bounded on $[-n, n]$. Moreover, $(P(t)h, h) \geqslant 0$ for all

$h\in H$, and $r\in L^2(-n,n;H)$. The Faedo-Galerkin approximation of the problem (3.6), (3.7), (3.9) has the same form: it is necessary to replace V and H by the finite-dimensional spaces $V_m=H_m$:

$$\frac{du_{1m}}{dt}+A_m(t)u_{1m}+u_{2m} = 0,$$

$$-\frac{du_{2m}}{dt}+A_m^*(t)u_{2m}-u_{1m} = f_{2m},$$

$$u_{1m}(-n) = 0,\ u_{2m}(n) = 0.$$

Moreover, estimate (3.5) holds for A_m with the same constant α, and $\lim f_{2m}=f_2$ in $L^2(-n,n;V')$. This problem admits decoupling (similar to the one above). Moreover, $\lim r_m=r$ weakly in $L^2(-n,n;H)$ and $\lim P_m(t)=P(t)$ in a suitable sense, which we need not specify.

The function r_m is a solution of the problem

$$-\frac{dr_m}{dt}+A_m^*r_m+P_mr_m = f_{2m},\ r_m(n) = 0$$

(see [45, Ch. 4, (4.69)]. We multiply this equality by r_m and then integrate. Taking into account that the operators $P_m(t)$ are nonnegative, we obtain with the aid of (3.5) the inequality

$$\frac{1}{2}|r_m|^2\Big|_{t_1}^{t_2}+\alpha\int_{t_1}^{t_2}||r_m||^2dt \leqslant \int_{t_1}^{t_2}||r_m||\cdot||f_{2m}||_*dt.$$

Reversing the time and using Lemma 1.1 of Ch. 2 we obtain an estimate

$$||r_m||_{S^2} \leqslant C||f_{2m}||_{S^2},$$

where C does not depend on m and n. Passing to the limit when $m\to\infty$ we have

$$||r||_{S^2} \leqslant C||f_2||_{S^2}. \tag{3.11}$$

Here the S^2-norm on the right (left) is that of functions with values in V' (respectively, V). Thus we can interchange V and V'.

By (3.6), (3.10), u is a solution of the problem

$$\frac{du}{dt}+Au_1+Pu_1 = -r,\ u_1(-n) = 0.$$

Therefore, as in the proof of (3.11),

$$||u_1||_{S^2} \leqslant C||r||_{S^2} \leqslant C||f_2||_{S^2}, \tag{3.12}$$

$$||u_1||_{C([-n,n];H)} \leqslant C||r||_{S^2} \leqslant C||f_2||_{S^2}. \tag{3.13}$$

Finally, (3.7) together with (3.12) and Lemma 1.1 (Ch. 2) gives the estimate

$$||u_2||_{S^2} \leqslant C||f_2||_{S^2},$$

$$||u_2||_{C([-n;n];H)} \leqslant C||f_2||_{S^2}.$$

Thus there is an estimate for the approximate solution u,

$$||u_{in}||_{S^2}+||u_{in}||_{C([-n;n];H)} \leqslant C||f_2||_{S^2}, \tag{3.14}$$

which is uniform with respect to n.

Taking a subsequence if necessary, we may assume that for arbitrary $a<b$ the limit $\lim u_{in}=u_i$ exist weakly in $L^2(a,b;V)$ and $\star$-weakly in $L^\infty(a,b;H)$. Furthermore, by virtue of equations (3.6), (3.7) the derivatives u_{in}' are bounded in $L^2(a,b,V')$. Therefore we may assume that $\lim u_{im}'=u_i'$ weakly in $L^2(a,b;V')$. It is easy to see that the limits satisfy equations (3.6), (3.7) and, by virtue of (3.14), lie in $BS^2(\mathbf{R};V)\cap C_b(\mathbf{R};H)$. Moreover, taking into account the reduction made in the beginning of the existence proof, there is an estimate for the bounded solution just constructed:

$$\sum_{i=1,2}(||u_i||_{S^2}+||u_i||_{C_b(\mathbf{R},H)}) \leqslant C(||f_1||_{S^2}+||f_2||_{S^2}), \tag{3.15}$$

where C depends on α only.

3) *Almost periodicity*. For the problem shifted by $s\in\mathbf{R}_B$,

$$\frac{du_1}{dt}+A(t-s)u_1+u_2 = f_1(t-s), \tag{3.6s}$$

$$-\frac{du_2}{dt}+A^*(t-s)u_2-u_1 = f_2(t-s), \tag{3.7s}$$

there exists a unique bounded solution $u_s=(u_{1s}, u_{2s})$. Obviously, $u_s(t)=u(t-s)$ for $s\in\mathbf{R}$. The difference $v=u_{s_0}-u_s$ satisfies equations of the form (3.6s), (3.7s), where the right-hand sides are replaced by

$$g_1(t) = f_1(t-s_0)-f_1(t-s)-[A(t-s_0)-A(t-s)]u_{1s}(t),$$

$$g_2(t) = f_2(t-s_0)-f_1(t-s)-[A^*(t-s_0)-A^*(t-s)]u_{2s}(t).$$

Obviously $\lim_{s\to s_0} g_i=0$ in $BS^2(\mathbf{R}; V')$. Therefore estimate (3.15) shows that u_s depends continuously on $s\in\mathbf{R}_B$ in $BS^2(\mathbf{R}; V)\bigcap C_b(\mathbf{R}; H)$. Thus almost periodicity is established. □

REMARK 3.4. It is easy to see that the uniqueness proof may be extended to the case of the nonlinear system (3.3), (3.4) if we require estimates of the type (1.20) to hold for the operators A_i. The remaining arguments from the proof of Theorem 3.3 are essentially linear. Note that under certain additional restrictions the function $r(t)$ also satisfies the inverse parabolic equation similar to that written above for $r_m(t)$ (see [45, Ch. 3, Theorem 4.4]). Here we have used the Faedo-Galerkin approximation in order to avoid superfluous restrictions.

REMARK 3.5. The existence of estimate (3.15) for bounded solutions allows us to obtain the following statement about regularity of solutions with respect to t. Here we use the same arguments as in the proof of point 3) of Theorem 3.3. Let

$$||A(t)-A(t-s)|| \leqslant C\min(1, s^\gamma),\ 0<\gamma\leqslant 1,$$

and $f\in BS_r^{2,\theta}(\mathbf{R}; V')$, where $0<\theta<\gamma$ if $r<\infty$ and $0<\theta\leqslant\gamma$ if $r=\infty$. Then $u_i\in BS_r^{2,\theta}(\mathbf{R}; V)\bigcap C_{b,r}^{\theta}(\mathbf{R}; H)$.

REMARK 3.6. Similarly to the proof of Theorem 3.3 we can establish the following. Let $\xi\in H$ and $f_i\in BS^2(\mathbf{R}_+; V')$. Then the system (3.6), (3.7) has a unique solution (u_1, u_2) with $u_i\in BS^2(\mathbf{R}_+; v)\bigcap C_b(\mathbf{R}_+; H)$ and $u_1(0)=\xi$. This assertion, as well as the previous results, is of some interest in optimal control theory [45, Ch. 3, § 6].

3.2. CERTAIN DEGENERATE PROBLEMS.

Consider the problem ($1<p<\infty$):

$$x\frac{du}{dt}+(-1)^m\frac{\partial^m}{\partial x^m}\left[\left|\frac{\partial^m u}{\partial x^m}\right|^{p-2}\cdot\frac{\partial^m u}{\partial x^m}\right] = f(t,x),$$

$$(t,x)\in\mathbf{R}\times(-1,1), \tag{3.16}$$

$$\frac{\partial^k u}{\partial x^k}(t,\pm 1) = 0,\ 0\leqslant k\leqslant m-1. \tag{3.17}$$

The analogous problem with boundary conditions with respect to time was considered in [104, 44, Ch. 3, § 2]. We will show that it is easy to establish the existence of Besicovitch a.p. solutions for this problem. On the contrary, it is not quite clear how Stepanov a.p.solutions can be studied.

Let $\mathcal{V}=B^p(\mathbf{R}; W_0^{m,p}(-1,1))$, then $\mathcal{V}'=B^{p'}(\mathbf{R}; W^{-m,p'}(-1,1))$. Put

$$D(\Lambda) = \{v\in S^p(\mathbf{R}; W_0^{m,p}(-1,1))(\mid x\frac{\partial v}{\partial t}\in S^{p'}(\mathbf{R}; W^{-m,p'}(-1,1))\}.$$

On $D(\Lambda)$ we define the operators

$$\Lambda v = x\frac{\partial v}{\partial t},$$

$$\Lambda^+ v = -x\frac{\partial v}{\partial t}.$$

We have

$$(\Lambda v, w)_B = (v, \Lambda^+ w)_B,\ v, w\in D(\Lambda). \tag{3.18}$$

Indeed, from Proposition 2.1 [44, Ch. 3] it follows that $|x|^{1/2}\cdot v(t,x)\in \mathrm{CAP}(\mathbf{R}; L^2(-1,1))$ and

$$\int_{-T}^{T}\int_{-1}^{1} x\cdot v'w\,dt\,dx = -\int_{-T}^{T}\int_{-1}^{1} v\cdot xw'\,dt\,dx + \int_{-1}^{1} x\cdot v\cdot w\,dx\,\Big|_{-T}^{T}.$$

If we divide this equality by $2T$ and pass to the limit as $T\to\infty$ we obtain (3.18).

Obviously $D(\Lambda)$ is dense in $\mathcal{V}$. From (3.18) it follows that Λ admits a closure as an operator from $\mathcal{V}$ to $\mathcal{V}'$. This closure will be denoted by L. Obviously $L^* \supset \Lambda^+$. The Bochner-Fejér operators with respect to the variable t commute with L, and hence with L^*. This implies that L^* is a closure of Λ^+. Thus $D(L)=D(L^*)$ and $L^*=-L$. Moreover, L is a maximal monotone operator. If we interpret problem (3.16), (3.17) as an equation

$$Lu+\mathbf{A}u = f, \tag{3.19}$$

where $\mathbf{A}:\mathcal{V}\to\mathcal{V}'$ is a bounded monotone semicontinuous and coercive operator, we obtain following result:

THEOREM 3.7. *For an arbitrary* $f\in B^{p'}(\mathbf{R};W^{-m,p'}(-1,1))$ *problem* (3.16), (3.17) *has a unique solution* $u\in B^{p}(\mathbf{R};W_0^{m,p}(-1,1))$. □

Uniqueness follows from the strong monotonicity of the operator A which corresponds to the nonlinear term in (3.16). Of course, more general nonlinearities may be considered here.

As has been noted above, the question about the existence of Stepanov a.p. solutions remains open. However, consider instead of (3.16) the modified equation

$$|x|\frac{\partial u}{\partial t}+(-1)^m\frac{\partial^m}{\partial x^m}\left(\left|\frac{\partial^m u}{\partial x^m}\right|\frac{\partial^m u}{\partial x^m}\right) = f \tag{3.20}$$

with the same boundary conditions (3.17). There is an assertion analogous to Theorem 3.7 for this problem. Now let $p\geqslant 2$; consequently $W_0^{m,p}(-1,1)\subset L^2(-1,1)$. Consider a solution u_n of the problem (3.20), (3.17) on the half-line $[-n,+\infty)$ that satisfies the initial condition $u_n(-n)=0$. Modifying in an obvious manner the arguments of [44, Ch. 3, n°. 2.6], it is easy to see that such a solution exist in the space

$$L^p_{\mathrm{loc}}([-n;\infty);V)\cap C([-n,\infty),H_{1/2}),$$

where $V=W_0^{m,p}(-1,1)$, $H_{1/2}=\{v\,|\,|x|^{1/2}\cdot v\in L^2(-1,1)\}$ with the norm $|\cdot|_{1/2}$. Since for the operator $\mathbf{A}$ (the stationary part of the equations (3.16) and (3.20)) there is an estimate

$$(\mathbf{A}v-\mathbf{A}w,v-w) \geqslant \alpha\,\|v-w\|^p, \tag{3.21}$$

equation (3.20) multiplied by u implies

$$\frac{1}{2}|u_n|_{1/2}^2|_{t_1}^{t_2}+\alpha\int_{t_1}^{t_2}||u_n||^2dt \leqslant \int_{t_1}^{t_2}||u_n||\cdot||f||_*dt.$$

Lemma 1.1 of Ch. 2 implies that the u_n are uniformly bounded in $BS^p(\mathbf{R}; V)\cap C_b(\mathbf{R}; H_{1/2})$. Exactly in the same way there is an estimate (u_n is extended by zero to $\mathbf{R}$, f_n is the cut-off of f by zero to the left of the point $-n$ extended by zero)

$$\frac{1}{2}|u_n-u_m|_{1/2}^2|_{t_1}^{t_2}+\alpha\int_{t_1}^{t_2}||u_n-u_m||^pdt \leqslant \int_{t_1}^{t_2}||u_n-u_m|\cdot|f_n-f_m||_*dt.$$

As in n°. 1.2 this implies that $u_n\to u$ in $C(\mathbf{R}; H_{1/2})$ and weakly in $L^p_{loc}(\mathbf{R}; V)$; moreover $u\in BS^p(\mathbf{R}; V)\cap C_b(\mathbf{R}; H_{1/2})$. The passage to the limit is justified in the same way as in § 1. This leads to the following theorem.

THEOREM 3.8. *For an arbitrary* $f\in BS^{p'}(\mathbf{R}; W^{-m,p'}(-1,1))$ *the problem* (3.20), (3.16) *has a unique solution* $u\in BS^p(\mathbf{R}; W_0^{m,p}(-1,1))$ *such that* $|x|^{1/2}u\in C_b(\mathbf{R}; L^2(-1,1))$. *Moreover, if* $f\in S^{p'}(\mathbf{R}; W^{-m,p'}(-1,1))$, *then* $u\in S^p(\mathbf{R}; W_0^{m,p}(-1,1))$ *and* $|x|^{1/2}\cdot u\in CAP(\mathbf{R}; L^2(-1,1))$.

REMARK 3.9. Equation (3.16) is invariant under translation. Therefore, by estimate (3.21) the following regularity theorem holds for solutions of the problem (3.16), (3.17) (compare with Ch. 3, n°. 5.3)): If $p\geqslant 2$ and $f\in B_r^{p',\theta}(\mathbf{R}; W^{-m,p'}(-1,1))$ where $0<\theta<1$ if $r<\infty$ and $0<\theta\leqslant 1$ if $r=\infty$, then $u\in B_\nu^{p,\mu}(\mathbf{R}; W_0^{m,p}(-1,1))$, where $\mu=\theta/(p-1)$, $\nu=r(p-1)$. The same is true for the problem (3.20), (3.17). Furthermore, in the last case the fact that $f\in BS_r^{p',\theta}(\mathbf{R}; W^{-m,p'}(-1,1))$ implies that $u\in BS_l^{p,\eta}(\mathbf{R}; W_0^{m,p}(-1,1))$ and $|x|^{1/2}u\in C_{b,p}^{\mu}(\mathbf{R}; L^2(-1,1))$, where $l=rp(p-1)/2$, $\eta=2\theta/p(p-1)$. Note that for a problem (3.16), (3.17) with boundary conditions in time, considered in [44, 104], no assertion about regularity with respect to time is known, even not in case $p=2$ [44, Ch. 3, Problem 11.1].

REMARK 3.10. In Theorem 3.8 and in Remark 3.9 we may consider a $p<2$ for which $W_0^{m,p}(-1,1)\subset L^2(-1,1)$. In this case estimate (3.20) must be replaced by the estimate (3.2) of Ch. 3 with $E=W_0^{m,p}(-1,1)$ and $q=2$. Theorem 3.8 remains true, but changes in the spirit of Theorems 3.5 and 5.3 of Ch. 3 must be made in the assertion concerning regularity.

Comments

§§ 1, 2. These results were stated by the author [68]. They were partially announced in [66]. The Cauchy problem and the mixed problem for equations considered were investigated in [104] (see also [44]). Symmetric hyperbolic systems with monotone nonlinearities were considered for the first time in [134]. See also the paper [143], in which mixed problems for linear and nonlinear Schrödinger equations were studied.

§ 3. These results are published for the first time.

CHAPTER 5.

Problems that are almost periodic in space variables.

1. Nonlinear elliptic equations.

1.1. MONOTONICITY AND COERCIVITY OF NONLINEAR ALMOST PERIODIC OPERATORS.

Now we will consider differential operators of the form

$$A(u) = \sum_{|\alpha|=0}^{m} (-1)^{|\alpha|} \partial^{\alpha} A_{\alpha}(x, \delta^m u). \tag{1.1}$$

Here we use standard notations for multi-indices and derivatives, and $\delta^m = \{\partial^\alpha \mid 0 \leqslant |\alpha| \leqslant m\}$ is the collection of all partial derivatives of order not exceeding m (the number of such derivatives is denoted by M). For the real-valued functions $A_\alpha(x, \xi)$, $|\alpha| \leqslant m$, we assume that

$$|A_\alpha(x, \xi)| \leqslant c|\xi|^{p-1}, \ (x, \xi) \in \mathbf{R}^n \times \mathbf{R}^M, \tag{1.2}$$

for some $p>1$ and $c>0$. We also assume that the $A(x, \xi)$ are continuous in $\xi \in \mathbf{R}^M$ for any fixed $x \in \mathbf{R}^n$ and that the family $\{|\xi|^{1-p} A_\alpha(x, \xi) \mid \xi \in \mathbf{R}^M\}$ is equicontinuous and uniformly a.p. in $x \in \mathbf{R}^n$. The functions $A_\alpha(x\, \xi)$ can be continuously extended to functions defined for $x \in \mathbf{R}^n_B$, $\xi \in \mathbf{R}^n$ (for such extensions we use the same notations). Moreover, estimate (1.2) remains valid for all $(x, \xi) \in \mathbf{R}^n_B \times \mathbf{R}^M$.

The operator (1.1) may be regarded as a continuous operator $A: W^{m,p}(\mathbf{R}^n) \rightarrow W^{-m,p'}(\mathbf{R}^n)$, and

$$\| A(u) \|_{-m,p} \leqslant c \| u \|_{m,p}^{p-1}, \ u \in W^{m,p}(\mathbf{R}^n) \tag{1.3}$$

(see, for example, [44]). Since the A_α clearly satisfy the Carathéodory condition on $\mathbf{R}^n_B$, (1.1) defines also a continuous operator $A_B: W^{m,p}(\mathbf{R}^n_B) \to W^{-m,p'}(\mathbf{R}^n_B)$ (see Ch. 2, Remark 2.7), for which the estimate

$$\| A_B(u) \|_{-m,p',B} \leqslant c \| u \|^{p-1}_{m,p,B},\ u \in W^{m,p}(\mathbf{R}^n_B), \tag{1.4}$$

is valid.

Now we will study relations between properties of A and A_B which are essential for the monotonicity method. We make use of the following terminology. An operator $A: V \to V'$ is said to be *p-coercive* if

$$(A(u), u) \geqslant \alpha \| u \|^p,\ \alpha > 0,\ u \in V;$$

p-monotone if

$$(A(u) - A(v), u - v) \geqslant \alpha \| u - v \|^p,\ \alpha > 0,\ u, v \in V.$$

THEOREM 1.1. *The following assertions hold*:
a) A is a monotone operator iff A_B is a monotone operator;
b) A is p-coercive iff A_B is p- coercive;
c) A is p-monotone iff A_B is p- monotone.

REMARK 1.2. This result seems to be similar to a result in [90] on coincidence of spectra for linear a.p. operators on L^2 and B^2. It shows that the monotonicity method can be applied to both operators A and A_B simultaneously. We note that with strict monotonicity the corresponding asserting does not hold. As an example we take the Laplace operator. It is obvious that $-\Delta: H^1(\mathbf{R}^n) \to H^{-1}(\mathbf{R}^n)$ is a strictly monotone operator, but $-\Delta_B: H^1(\mathbf{R}^n_B) \to H^{-1}(\mathbf{R}^n_B)$ is not $(-\Delta_B 1 = 0)$.

We now state an auxiliary result, similar to Lemma 4.1 in [91]. Consider a non-negative function $\psi_R \in C_0^\infty(\mathbf{R}^n)$, $R \geqslant 1$, such that for some $\kappa \in (0, 1)$,

$$\psi_R(x) = \begin{cases} 1, & |x| \leqslant R, \\ 0, & |x| \geqslant R + R^\kappa, \end{cases}$$

$$| \partial^\gamma \psi_R(x) | \leqslant C_\gamma \cdot R^{-\kappa|\gamma|},$$

where $C_\gamma > 0$ does not depend on R. By the definition of mean value we see that

$$M\{f\} = \lim_{R\to\infty} \frac{1}{\omega_R} \int_{V_R} \psi_R(x) f(x)dx, \ f \in \mathrm{CAP}(\mathbf{R}^n), \tag{1.5}$$

where $V_R = \{x \in \mathbf{R}^n : |x| \leqslant R\}$ is the R-ball centred at 0 and $\omega_R = \text{mes}\, V_R$.

LEMMA 1.3. *For any* $u, v \in \mathrm{CAP}^\infty(\mathbf{R}^n)$ *we have*

$$||u||^p_{m,p,B} = \lim_{R\to\infty} \frac{1}{\omega_R} ||\psi_R \cdot u||^p_{m,p}, \tag{1.6}$$

$$(A(u), v)_B = \lim_{R\to\infty} \frac{1}{\omega_R} (A(\psi_R \cdot u), \psi_R \cdot v). \tag{1.7}$$

Proof. First we prove the equality

$$||\partial^\alpha u||^p_{B^p} = \lim \frac{1}{\omega_R} ||\partial^\alpha(\psi_R u)||^p_{L^p}, \tag{1.8}$$

implying (1.6). By Leibniz' formula we have

$$\partial^\alpha(\psi_R u) = \psi_R \cdot \partial^\alpha u + \sum_{\substack{\beta+\gamma=\alpha \\ \beta\neq 0}} ||\partial^\alpha \psi_R \cdot \partial^\alpha u||_{L^p} \leqslant ||\partial^\alpha(\psi_R u)||_{L^p} \leqslant$$

$$\leqslant ||\psi_R \partial^\alpha u||_{L^p} + \sum c_{\beta,\gamma} ||\partial^\alpha \psi_R \cdot \partial^\gamma u||_{L^p}.$$

Since the $\psi_R^p(x)$ are uniformly bounded, $\psi_R^p(x) = 1$ for $|x| \leqslant R$ and $\psi_R^p(x) = 0$ for $|x| \geqslant R + R^\kappa$ we have

$$||\partial^\alpha u||^p_{B^p} = M\{|\partial^\alpha u|^p\} = \lim \frac{1}{\omega_R} \int \psi_R^p(x) |\partial^\alpha u(x)|^p dx =$$

$$= \lim \frac{1}{\omega_R} ||\psi_R \cdot \partial^\alpha u||^p_{L^p}$$

(compare with (1.5)). For $\beta \neq 0$ we have

$$\int |\partial^\beta \psi_R(x)|^p |f(x)| dx = \int_{S_R} |\partial^\beta \psi_R(x)|^p |f(x)| dx \leqslant$$

$$\leqslant C_\beta^p \cdot R^{-\kappa|\beta|_p} \cdot \mathrm{mes}\, S_R,$$

where $S_R = V_{R+R^\kappa} \setminus V_R$. The measure of S_R is equal to $c \cdot R^{n-1+\kappa}$. Since $\omega_R = cR^n$, our integral is $o(\omega_R)$. Hence

$$\lim \frac{1}{\omega_R} \| \partial^\gamma \psi_R \cdot \partial^\gamma u \|_{L^p}^p = 0$$

and (1.8) is proved.

Now we show that

$$(A(u), v)_B = \lim_{R\to\infty} \frac{1}{\omega_R} (\psi_R \cdot A(u), \psi_R \cdot v). \tag{1.9}$$

Indeed,

$$(\psi_R \cdot A(u), \psi_R \cdot v) = (A(u), \psi_R^2 \cdot v) = \sum_{|\alpha| \leqslant m} \int A_\alpha(x, \delta^m u(x)) \partial^\alpha (\psi_R^2(x) \cdot v(x)) dx.$$

Since the properties of ψ_R^2 are similar to those of ψ_R, (1.9) may be proved in the same way as (1.8).

Now we consider

$$I_R = (\psi_R \cdot A(u) - A(\psi_R \cdot u), \psi_R \cdot v).$$

It is easy to see that

$$| I_R | \leqslant \sum_{|\alpha| \leqslant m} \left[\int_{S_R} | A_\alpha(x, \delta^m u(x)) | \cdot | \partial^\alpha (\psi_R^2(x) v(x)) | dx + \right.$$

$$\left. + \int_{S_R} | A_\alpha(x, \delta^m (\psi_R(x) u(x)) | \cdot | \partial^\alpha (\psi_R(x) v(x)) | dx \right].$$

Inequality (1.2) and the properties of ψ_R imply that the integrands are uniformly bounded in R. Hence

$$| I_R | \leqslant C \mathrm{mes}\, S_R = C_1 R^{n-1} \cdot R^\kappa.$$

Thus

$$\omega_R^{-1} \cdot I_R = o(R^{\kappa-1}),$$

and (1.9) implies (1.7). □

Proof. of Theorem 1.1. By the continuity of A_B it is sufficient to prove all inequalities we need only for $\mathrm{CAP}^\infty(\mathbf{R}^n)$ functions. But in this case the inequalities follow from corresponding inequalities for A and Lemma 1.3. Hence all the implications $\Rightarrow$ of Theorem 1.1 are proved.

Now we prove the implications $\Leftarrow$. We will only consider assertion c), because the other assertion can be treated in a similar way. First of all, by applying the Bochner-Fejér operators to $A_\alpha(x, \xi)$ we find trigonometric polynomials $A_\alpha^\epsilon(x, \xi)$ in x, with coefficients depending on ξ, such that

$$| A_\alpha(x, \xi) - A_\alpha^\epsilon(x, \xi) | \leqslant \epsilon | \xi |^{p-1}, \ (x, \xi) \in \mathbf{R}^n \times \mathbf{R}^M.$$

Setting

$$A^\epsilon(u) = \sum_{|\alpha| \leqslant m} (-1)^{|\alpha|} \partial^\alpha A_\alpha^\epsilon(x, \delta^m u),$$

we have

$$|| A(u) - A^\epsilon(u) ||_{-m,p'} \leqslant \epsilon || u ||_{m,p}^{p-1}, \ u \in W^{m,p}(\mathbf{R}^n), \tag{1.10}$$

$$|| A_B(h) - A_B^\epsilon(h) ||_{-m,p',B} \leqslant \epsilon || h ||_{m,p,B}^{p-1}, \ h \in W^{m,p}(\mathbf{R}_B^n). \tag{1.11}$$

Now for a fixed $\epsilon > 0$ the spectra of A_α^ϵ generate a k-dimensional vector space over $\mathbf{Q}$, $k < \infty$. Hence there is an embedding $j: \mathbf{R}^n \subset \mathbf{R}^k$ such that A_α^ϵ may be extended to functions $\tilde{A}_\alpha^\epsilon$ on $\mathbf{R}^k$ which are periodic in $\{y_i\}$ ($\{y_i\}$ is the standard coordinate system in $\mathbf{R}^k$). Without loss of generality we may assume these $\tilde{A}_\alpha^\epsilon$ to be 1-periodic. Moreover, by the Kroneker-Weyl theorem, for any r-periodic ($r \in \mathbf{Z}_+$) function f on $\mathbf{R}^k$ we have

$$M_x\{j^* f\} = M_y\{f\}.$$

Let now $u \in C_0^\infty(\mathbf{R}^n)$. Identifying $\mathbf{R}^n$ with $j(\mathbf{R}^n)$ we introduce orthogonal coordinates (x, x') in $\mathbf{R}^k$ such that (x) are the original coordinates in $\mathbf{R}^n$. For $r \in \mathbf{Z}_+$ we consider an r-cube $K \subset \mathbf{R}^k$ with edges along the y-axes and such that the δ-neighbourhood (in $\mathbf{R}^k$) of $\operatorname{supp} u \subset \mathbf{R}^n \subset \mathbf{R}^k$ is contained in K for all sufficiently small $\delta<0$. Now we define an r-periodic (in y) function $\tilde{h}_{\epsilon,\delta}$ on $\mathbf{R}^k$ by setting

$$\tilde{h}_{\epsilon,\delta}(x,x') = \begin{cases} u(x), & (x,x')\in k, \quad |x'| \leqslant \delta, \\ 0, & (x,x')\in k, \quad |x'| > \delta. \end{cases}$$

Denote by $h_{\epsilon,\delta}$ its restrictions to $\mathbf{R}^n$, i.e. $h_{\epsilon,\delta}=j^*\tilde{h}_{\epsilon,\delta}$. This function is continuous on $\mathbf{R}^n$ (while for $\tilde{h}_{\epsilon,\delta}$ this is not true), and

$$h_{\epsilon,\delta}\in W^{m,p}(\mathbf{R}^n_B).$$

Indeed,

$$\partial^\alpha_x\tilde{h}_{\epsilon,\delta}\in B^p_{\mathbf{Z}^k/r}(\mathbf{R}^k) = L^p(\mathbf{R}^k/r\mathbf{Z}^k),$$

and we may apply the results of n°. 3.1 of Ch. 1. Hence

$$\partial^\alpha_x h_{\epsilon,\delta} = \partial^\alpha_x j^*\tilde{h}_{\epsilon,\delta} = j^*\partial^\alpha_x\tilde{h}_{\epsilon,\delta}\in B^p(\mathbf{R}^n).$$

Moreover, since j^* is isometric we have

$$||\,\partial^\alpha_x h_{\epsilon,\delta}\,||^p_{B^p} = M_y\{|\,\partial^\alpha_x\tilde{h}_{\epsilon,\delta}\,|^p\} = \frac{1}{r^k}\int_K|\,\partial^\alpha_x\tilde{h}_{\epsilon,\delta}\,|^p dx =$$

$$= \frac{\delta^{k-n}}{r^k}\int_K|\,\partial^\alpha_x u(x)\,|^p dx = \frac{\delta^{k-n}}{r^k}||\,\partial^\alpha_x u\,||^p_{L^p}.$$

Hence

$$||\,h_{\epsilon,\delta}\,||^p_{m,p,B} = \frac{\delta^{k-n}}{r^k}||\,u\,||^p_{m,p}. \tag{1.12}$$

Now we have

$$(A^\epsilon_B(h_{\epsilon,\delta}),h_{\epsilon,\delta}) = \sum_{|\alpha|\leqslant m} M_x\{A^\epsilon_\alpha(x;\delta^m_x h_{\epsilon,\delta})\cdot\partial^\alpha_x h_{\epsilon,\delta}\} =$$

$$= \frac{1}{r^k}\sum_{|\alpha|\leqslant m}\int_K\tilde{A}^\epsilon_\alpha(y,\delta^m_x\tilde{h}_{\epsilon,\delta})\cdot\partial^\alpha_x\tilde{h}_{\epsilon,\delta}dy =$$

$$= \frac{1}{r^k} \sum_{|\alpha| \leqslant m} \int_{|x'| \leqslant \delta} \int \Big[\tilde{A}^{\epsilon}_{\alpha}(x, 0, \delta^m_x u(x)) \partial^{\alpha}_x u(x) +$$

$$+ \Big[\tilde{A}^{\epsilon}_{\delta}(x, x', \delta^m_x u(x)) - \tilde{A}^{\epsilon}_{x}(x, 0, \delta^m_x u(x)) \Big] \partial^{\alpha}_x u(x) \Big] dx dx'.$$

Here the first summand is equal to

$$\frac{\delta^{k-n}}{r^k} \sum_{|\alpha| = m} \int \tilde{A}^{\epsilon}_{\alpha}(x, 0, \delta^m_x u(x)) \partial^{\alpha}_x u(x) dx =$$

$$= \frac{\delta^{k-n}}{r^k} \sum_{|\alpha| \leqslant m} \int A^{\epsilon}_{\alpha}(x, \delta^m_x u(x)) \partial^{\alpha}_x u(x) dx = \frac{\delta^{k-n}}{r^k} (A^{\epsilon}(u), u).$$

The functions $|\xi|^{1-p} \tilde{A}^{\epsilon}_{\alpha}(x, x', \xi)$ are equicontinuous. Hence for sufficiently small $\delta > 0$ we have

$$| \tilde{A}^{\epsilon}_{\alpha}(x, x', \delta^m_x u(x) - \tilde{A}^{\epsilon}_{\alpha}(x, 0, \delta^m_x u(x)) | \leqslant \epsilon \, |\delta^m_x u(x)|^{p-1}, \ |x'| \leqslant \delta.$$

Consequently,

$$| (A^{\epsilon}_B(h_{\epsilon,\delta}), h_{\epsilon,\delta})_B - \frac{\delta^{k-n}}{r^k} (A^{\epsilon}(u), u) | \leqslant \epsilon \frac{\delta^{k-n}}{r^k} \| u \|^p_{m-p}. \tag{1.13}$$

Now, by (1.10) we have

$$(A(u), u) \geqslant (A^{\epsilon}(u), u) - \epsilon \| u \|^p_{m,p}.$$

Using (1.13) we obtain

$$(A(u), u) \geqslant \frac{r^k}{\delta^{k-n}} (A^{\epsilon}_B(h_{\epsilon,\delta}), (h_{\epsilon,\delta})_B - 2\epsilon \| u \|^p_{m,p},$$

and by (1.11)

$$(A(u), u) \geqslant \frac{r^k}{\delta^{k-n}} (A_B(h_{\epsilon,\delta}, h_{\epsilon,\delta})_B - \epsilon \frac{r^k}{\delta^{k-n}} \| h_{\epsilon,\delta} \|^p_{m,p,B} - 2\epsilon \| u \|^p_{m,p}.$$

Since A_B is p-coercive, this implies

$$(A(u), u) \geqslant (\alpha-\epsilon)\frac{r^k}{\delta^{k-n}}\,||\, h_{\epsilon,\delta}\,||^p_{m,p,B}-2\epsilon\,||\, u\,||^p_{m,p}.$$

Taking into account (1.12) we obtain

$$(A(u), u) \geqslant (\alpha-3\epsilon)\,||\, u\,||^p_{m,p}.$$

Finally, since $\epsilon>0$ is arbitrarily small, we have

$$(A(u), u) \geqslant \alpha\,||\, u\,||^p_{m,p} \tag{1.14}$$

for any $u\in C_0^\infty(\mathbf{R}^n)$. In fact, by the continuity of A inequality (1.14) holds for any $u\in W^{m,p}(\mathbf{R}^n)$ and any A which is p-coercive. □

REMARK 1.4. In fact, the constants which appear in the p-coercive (p-monotonicity) estimates for the operators A and A_B may be taken equal.

1.2. SECOND ORDER ALMOST PERIODIC ELLIPTIC EQUATIONS.

The monotonicity method, if applicable, gives rise to weak (in a certain sense) solutions of the equation $A(u)=f$, with A defined by (1.1), which belong to the space $W^{m,p}(\mathbf{R}^n_B)$. For second order elliptic equations (not in divergence form) we will show that one may find classical bounded solutions which are almost periodic in the sense of Besicovitch. To this end we use the method of successive approximation (or the sub- and super-solutions method). For the sake of simplicity we consider here operators with CAP$^\infty$-coefficients. Let

$$A = A(x,\partial) = -\sum_{i,j=1}^{n} a_{ij}(x)\frac{\partial^2}{\partial x_i\partial x_i}+\sum_{j=1}^{n} b_j(x)\frac{\partial}{\partial x_j}+c(x) \tag{1.15}$$

be a uniformly elliptic operator with CAP$^\infty$-coefficients and let $c(x)\geqslant 0$. We assume that the operator A is invertible. By Theorem 5.4 of [94], invertibility here may take place in any space $L^2(\mathbf{R}^n)$, $B^2(\mathbf{R}^n)$, $C_b^\infty(\mathbf{R}^n)$, or CAP$^\infty(\mathbf{R}^n)$. By results of [50], A^{-1} may be regarded as a bounded operator on the space $C_b(\mathbf{R}^n)$. Moreover, it is clear that $c(x)\not\equiv 0$. By the formula

$$A_s(x,\partial) = A(x+s,\partial),\ S\in\mathbf{R}^n_B,$$

the operator (1.15) generates a family $\{A_s\}$ of uniformly elliptic operators, called the hull of A. All such operators are invertible [91], and $c_s(x) \geqslant 0$, $c_s(x) \not\equiv 0$.

LEMMA 1.5. *Let* $g \in C_b^\infty(\mathbf{R}^n)$ *and* $g(x) \geqslant 0$. *If*

$$f(x) = Au(x) + g(x)u(x) \geqslant 0, \; u \in C_b^\infty(\mathbf{R}^n),$$

then $u(x) \geqslant 0$.

Proof. Let

$$\alpha = \inf_{x \in \mathbf{R}^n} u(x) < 0,$$

and let $\{s_k\} \subset \mathbf{R}^n$ be a minimizing sequence. Since $\mathbf{R}_B^n$ is compact we may assume that $\lim s_k = s \in \mathbf{R}_B^n$. The families of functions $\{u(\cdot + s_k)\}$, $\{f(\cdot + s_k)\}$ and $\{g(\cdot + s_k)\}$ are bounded in $C_b^\infty(\mathbf{R}^n)$. Hence, by the Arzelà-Ascoli theorem, the limits

$$u_0(x) = \lim u(x + s_k),$$

$$f_0(x) = \lim f(x + s_k),$$

$$g_0(x) = \lim g(x + s_k)$$

exist in the space $C^\infty(\mathbf{R}^n)$. In fact, these limit functions belong to $C_b^\infty(\mathbf{R}^n)$. Moreover, $f_0(x) \geqslant 0$, $g_0(x) \geqslant 0$ and

$$A_s u_0 + g_0 u_0 = f_0.$$

At $x = 0$ the function $u_0(x)$ attains its minimum, equal to α. By the classical maximum principle, $u_0(x) \equiv \alpha$ and (1.16) implies

$$\alpha \cdot [c_s(x) + g_0(x)] = 0.$$

Since $g_0(x) \geqslant 0$, $c_s(x) \geqslant 0$ and $c_s(x) \not\equiv 0$, this contradicts the negativity of α. □

LEMMA 1.6. *Let* $g \in C_b^\infty(\mathbf{R}^n)$, $g(x) \geqslant 0$ *and* $Au + gu = f$ *with* $u \in C_b^\infty(\mathbf{R}^n)$. *Then*

$$||u||_{C_b} \leqslant C||f||_{C_b}, \tag{1.17}$$

where $c>0$ does not depend on g and f.

Proof. Set $N=||f||_{C_b}$. Let $u_0\in C_b^\infty(\mathbf{R}^n)$ be a solution of $Au_0=N$ (it exists since A is integrable). By Lemma 1.5, $u_0(x)\geqslant 0$, and since $g(x)\geqslant 0$ we have

$$Au_0(x)+g(x)u_0(x) \geqslant N \geqslant f(x).$$

By Lemma 1.5 we see that $u(x)\leqslant u_0(x)$. Similarly, $u(x)\geqslant -u_0(x)$ and, therefore,

$$|u(x)| \leqslant u_0(x).$$

Since A^{-1} is bounded on $C_b(\mathbf{R}^n)$ we have

$$||u_0||_{C_b} \leqslant ||A^{-1}||N = ||A^{-1}||\,||f||_{C_b}.$$

This implies (1.17) with $C=||A^{-1}||$. □

Now we consider the equation

$$Au = f(x,u) \tag{1.18}$$

We assume that for all $\alpha\in\mathbf{Z}_+^{n+1}$ all functions $\partial^\alpha_{(x,u)}f(x,u)$ are a.p. on $x\in\mathbf{R}^n$, uniformly on any compact subset of values of $u\in\mathbf{R}$.

THEOREM 1.7. *Assume that $\partial f/\partial u(x,u)\geqslant 0$ and there are $\underline{u}, \overline{u}\in\mathrm{CAP}^\infty(\mathbf{R}^n)$ such that $\underline{u}(x)\leqslant\overline{u}(x)$ and*

$$A\overline{u}(x) \geqslant f(x,\overline{u}(x)), \tag{1.19}$$

$$A\underline{u}(x) \leqslant f(x,\underline{u}(x)).$$

Then there is a solution $u\in C_b^\infty(\mathbf{R}^n)$ of (1.18) such that $\underline{u}(x)\leqslant u(x)\leqslant\overline{u}(x)$ and $u\in B^\infty(\mathbf{R}^n)$.

Proof. We define a sequence $\{u_k\}$ of successive approximations by setting $u_0(x)=\overline{u}(x)$ and

$$A_{u_{k+1}}(x) = f(x,u_k(x)). \tag{1.21}$$

Since A is invertible, the sequence $\{u_k\}\subset\mathrm{CAP}^\infty(\mathbf{R}^n)$ is uniquely determined. Inequalities (1.19) and (1.20) imply

$$\underline{u}(x) \leqslant u_{k+1}(x) \leqslant u_k(x) \leqslant \bar{u}(x).$$

In particular, the sequence $\{u_k\}$ is bounded in the space $C_b(\mathbf{R}^n)$. Moreover, (1.21) and the results of [50] on S^p-solvability for elliptic equations show that $\{u_k\}$ is bounded in the space $C_b^\infty(\mathbf{R}^n)$. By the Arzelà-Ascoli theorem, $\{u_k\}$ is precompact in $C^\infty(\mathbf{R}^n)$, while the monotonicity of $\{u_k\}$ implies that $\lim u_k = u$ exists in this space. Obviously, $u \in C_b^\infty(\mathbf{R}^n)$ and it satisfies equation (1.18).

If we regard u_k as a function on $\mathbf{R}_B^n$ we obtain a decreasing sequence of functions on $\mathbf{R}_B^n$. The boundedness of $\{u_k\}$ implies that it converges pointwise on $\mathbf{R}_B^n$. The limit function $\tilde{u}$ is measurable and bounded on $\mathbf{R}_B^n$, i.e. $\tilde{u} \in B^\infty(\mathbf{R}^n)$. It is obvious that $\tilde{u}|_{\mathbf{R}^n} = u$. Moreover, for any $s \in \mathbf{R}_B^n$ the function $\tilde{u}|_{s+\mathbf{R}^n}$ is a C_b^∞-solution of the equation $A_s = u = f(x+s, u)$, and the theorem is proved. □

REMARK 1.8. The solution constructed in Theorem 1.7 is a function of the first Baire class on $\mathbf{R}_B^n$. Using results of [93] it is not difficult to obtain the following assertion. If there is uniqueness of C_b^∞-solutions of equation (1.18) lying between $\underline{u}$ and $\bar{u}$, then the solution constructed in Theorem 1.7 is a.p. in the sense of Levitan. Moreover, if there is uniqueness for all equations of the form

$$A_s u = f_s(x, u), \; s \in \mathbf{R}_B^n, \tag{1.22}$$

where $f_s(x, u) = f(x+s, u)$, then the solution constructed in Theorem 1.7 belongs to the space $\mathrm{CAP}^\infty(\mathbf{R}^n)$.

REMARK 1.9. The solution constructed in Theorem 1.7 is a maximal solution between $\underline{u}$ and $\bar{u}$. Using successive approximations with $u_0(x) = \underline{u}(x)$ we find that there is a minimal C_b^∞-solution, and it belongs to the space $B^\infty(\mathbf{R}^n)$.

As an application of Theorem 1.7 we now prove the following:

THEOREM 1.10 *Let $f(x, u) = f_1(x, u) + f_2(x, u)$, where $f_i(x, u)$, $i = 1, 2$, satisfy the same smoothness and almost periodicity conditions as f does. Assume that*

$$|f_1(x, u)| \leqslant C, \tag{1.23}$$

$$\frac{\partial f_2}{\partial u}(x, u) \leqslant 0, \tag{1.24}$$

for any $(x, u) \in \mathbf{R}^{n+1}$. Then equation (1.18) has a solution $u \in C_b^\infty(\mathbf{R}^n)$ such that $u \in B^\infty(\mathbf{R}^n)$.

Proof. a) Reduction to the case $f_2 \equiv 0$. Without loss of generality we may assume that $f_2(x, 0)=0$. For a function $\phi(x, u)$ we set

$$\bar{\phi}(x, u, v) = \int_0^1 \phi(x, tu+(1-t)v)dt. \tag{1.25}$$

If $u \in C_b^\infty(\mathbf{R}^n)$ is a solution of (1.18), then

$$Au+g(x)u = f_1(x, u),$$

where

$$g(x) = \overline{\left[\frac{\partial f_2}{\partial u}\right]}(x, u(x), 0) \in C_b^\infty(\mathbf{R}^n).$$

It is obvious that $g(x) \geqslant 0$ and, by Lemma 1.6,

$$\| u \|_{C_b} \leqslant C_1, \tag{1.26}$$

where $C_1>0$ depends only on the constant in inequality (1.23). Now we choose a smooth increasing function $\psi(t)$ such that $\psi(t)=t$ for $|t| \leqslant C_1$ and $\psi(t)=2C_1 \operatorname{sgn} t$ for $|t| \geqslant 2C_1$. We set

$$\tilde{f}(x, u) = f_1(x, u)+f_2(x, \psi(u)).$$

For equation (1.18) with f replaced by $\tilde{f}$ the a priori bound (1.26) is still valid with the same constant C_1. Hence, any bounded solution of this equation modified in this way is a bounded solution of (1.18), and the converse is valid too. Therefore we may assume that $f_2 \equiv 0$.

b). The case $f=f_1$. Set

$$\theta = \sup_{\substack{x \in \mathbf{R}^n \\ |u| \leqslant C_1}} \left| \frac{\partial f}{\partial u}(x, u) \right|,$$

where C_1 is the constant in estimate (1.26). Obviously, we may assume that $C \leqslant C_1$. The function $f(x, u)+\theta u$ is increasing for $|u| \leqslant C_1$. Clearly, equation (1.18) is equivalent to the equation

$$Au+\theta u = f(x,u)+\theta u. \tag{1.27}$$

We define $\overline{u}\in\mathrm{CAP}^\infty(\mathbf{R}^n)$ as the solution of the equation $A\overline{u}=C$ (such a solution exists and is unique) and set $\underline{u}=-\overline{u}$. By Lemma 1.5 we see that $\overline{u}\geqslant 0$. Moreover,

$$A\overline{u}+\theta\overline{u} = C+\theta\overline{u} \geqslant f(x,\overline{u})+\theta\overline{u},$$

$$A\underline{u}+\theta\underline{u} = -C+\theta\overline{u} \leqslant f(x,\underline{u})+\theta\underline{u}.$$

Hence equation (1.27) satisfies all the conditions of Theorem 1.7, and the required solution exists. □

REMARK 1.11. Remark 1.8 is still valid when applied to Theorem 1.10. We also note that in case $\partial f/\partial u\leqslant 0$ (i.e. $f=f_2$), a bounded solution of (1.18) is unique. Indeed, let $u, v\in C_b^\infty(\mathbf{R}^n)$ by two solutions of (1.18). Then for $w=u-v$ we have

$$Aw+h(x)w = 0,$$

where

$$h(x) = -\overline{\left[\frac{\partial f}{\partial u}\right]}(x,u(x),v(x))\in C_b^\infty(\mathbf{R}^n).$$

Since $h(x)\geqslant 0$, Lemma 1.6 implies $w=0$. Hence, in case $\partial f/\partial u\leqslant 0$ Theorem 1.10 gives the unique solution $u\in\mathrm{CAP}^\infty(\mathbf{R}^n)$ (since $\partial f_s/\partial u\leqslant 0$ for any $s\in\mathbf{R}_B^n$).

2. Almost periodic first order systems.

2.1. STATIONARY CASE.

We denote by APS^m, $m\in\mathbf{R}$, the class of a.p. symbols $a(x,\xi)$, $(x,\xi)\in\mathbf{R}^{2n}$, and by APL^m the corresponding class of pseudodifferential operators

$$a(x,D)u(x) = (2\pi)^{-n}\int e^{i(x-y)\xi}a(x,\xi)u(y)dyd\xi. \tag{2.1}$$

See [91] for details. Here we only note that the integral in (2.1) is regarded as oscillatory. Operators of such kind act naturally in scales $H^s(\mathbf{R}^n)$ and $H^s(\mathbf{R}_B^n)$.

Now we regard an $(N\times N)$-matrix of first order pseudodifferential operators, $A=a(x,D)$, as an unbounded operator in the space $B^2(\mathbf{R}^n)$ with domain

$$D(A) = \{u\in B^2(\mathbf{R}^n)\,|\,Au\in B^2(\mathbf{R}^n)\}$$

(The operator A acts on functions with values in $\mathbf{C}^N$, but for the sake of simplicity we will use the same notations for the functions spaces as in the scalar case.) It is a closed operator and it is a weak extension of the operator $A(x,D)$, which was initially defined on the space $\mathrm{CAP}^\infty(\mathbf{R}^n)$.

We begin with some auxiliary assertions on Friedrichs averaging operators in an a.p. situation. They are similar to the assertion of n°. 2.1. of Ch. 4 concerning partially a.p. averaging operators and may be proved by the same techniques. However, here we use another way to obtain the results, a way more suitable in this case.

Let $\phi\in C_0^\infty(\mathbf{R}^n)$ be a nonnegative function such that $\phi(x)=0$ for $|x|\geqslant 1$ and

$$\int\phi(x)dx = 1.$$

Set $\phi_\epsilon(x)=\epsilon^{-n}\phi(\epsilon^{-1}x)$ and

$$\Phi_\epsilon u = \phi_\epsilon * u = \phi_\epsilon(D)u.$$

Clearly $\Phi_\epsilon\in\mathrm{APL}^{-\infty}$.

LEMMA 2.1. *a)* $\Phi_\epsilon B^2(\mathbf{R}^n)\subset H^\infty(\mathbf{R}^n_B)$

b) for $u\in B^2(\mathbf{R}^n)$ *the inequality*

$$\|\Phi_\epsilon u\|_{B^2} \leqslant \|u\|_{B^2}$$

holds;

c) if $u\in B^2(\mathbf{R}^n)$, *then* $\lim_{\epsilon\to 0}\|\Phi_\epsilon u-u\|_{B^2}=0$.

Proof. Since $\Phi_\epsilon\in\mathrm{APL}^\infty$, a) is a simple consequence of general theorems on the action of a.p. pseudodifferential operators [91].

The theorem on coincidence of B^2- and L^2-norms of pseudodifferential operators [91] and well-known L^2-estimates for Φ_ϵ ([47]) imply b).

Since the norms of the Φ_ϵ are uniformly bounded, it suffices to prove c) for a subspace $\mathrm{CAP}(\mathbf{R}^n)$ which is dense in $B^2(\mathbf{R}^n)$. But since $\{\phi_\epsilon\}$ is δ-shaped, we have $\lim\Phi_\epsilon u=u$ in the space $C_b(\mathbf{R}^n)$ and, as a consequence, in $B^2(\mathbf{R}^n)$. □

Now let $b(x,\xi)\in \mathrm{APS}^0$ be a positive homogeneous symbol of order 0. Consider the commutators

$$T_\epsilon = [\Phi_\epsilon, b(x,D)\frac{\partial}{\partial x_j}].$$

LEMMA 2.2. *a) The operators* T_ϵ *are uniformly bounded on the space* $B^2(\mathbf{R}^n)$*;*

b) $\lim_{\epsilon\to 0} || T_\epsilon u ||_{B^2} = 0$ *for any* $u \in B^2(\mathbf{R}^n)$.

Proof. a) is a consequence of corresponding L^2-result [1] (see also [47] for the case $b(x,\xi)=b(x)$) and the theorem on coincidence of norms. Hence in the proof of b) we may assume that $u\in \mathrm{CAP}^\infty(\mathbf{R}^n)$.

Now recall that for any $v\in L^2(\mathbf{R}^n)$ the inequality

$$|| T_\epsilon u ||_{L^2} \leqslant C\cdot \sup_{|z|\leqslant\epsilon} || v(\cdot+z)-v(\cdot) ||_{L^2} \tag{2.2}$$

holds with $c>0$ depending only on $b(x,\xi)$ and ϕ (see [1] and, for the case $b(x,\xi)\equiv b(x)$, [47]).

Let ψ_R be the functions used in Lemma 1.3. Take in (2.2) the function $v(x)=\psi_R(x)u(x)$, where $u\in \mathrm{CAP}^\infty(\mathbf{R}^n)$. Then, using the same notations as in Lemma 1.3, we obtain

$$\frac{1}{\omega_R} || T_\epsilon(\psi_R\cdot u) ||_{L^2}^2 \leqslant$$

$$\leqslant \frac{C^2}{\omega_R}\cdot \sup_{|z|\leqslant\epsilon} || \psi_R(\cdot+z)u(\cdot+z)-\psi_R(\cdot)u(\cdot) ||_{L^2}^2. \tag{2.3}$$

Now we clearly have

$$|| \psi_R(\cdot+z)u(\cdot+z)-\psi_R(\cdot)u(\cdot) ||_{L^2}^2 = || \psi_R(\cdot+z)u(\cdot+z) ||_{l^2}^2 +$$

$$+ || \psi_R u ||_{L^2}^2 - 2\mathrm{Re}(\psi_R(\cdot+z)u(\cdot+z), \psi_R(\cdot)u(\cdot)).$$

By (1.5),

$$\lim_{R\to\infty}\frac{1}{\omega_R} || \psi_R(\cdot+z)u(\cdot+z) ||_{L^2}^2 = || u(\cdot+z) ||_{B^2}^2. \tag{2.5}$$

In fact, all functions in (2.5) are independent of z.

Now we prove that

$$\mathbf{M}\{f\} = \lim \frac{1}{\omega_R} \int \psi_R(x+z)\psi_R(x) f(x)dx, \tag{2.6}$$

where the limit exists uniformly in $z \in G$ for any compactum $G \subset \mathbf{R}^n$. Set

$$K_{z,R} = \{x \in \mathbf{R}^n \mid | z+x | \leqslant R, \ | x | \leqslant R\},$$

$$\omega_{z,R} = \operatorname{mes} K_{z,R}.$$

To compute $\omega_{z,R}$ we may assume that $z = (| z |, 0, \ldots, 0)$. Then we have

$$\omega_R \geqslant \omega_{z,R} \geqslant \operatorname{mes}(V_R \cap \{x \mid | x_1 | \geqslant | z | /2\}) \geqslant \omega_R - c_1 | z | R^{n-1}.$$

This implies that

$$\lim_{R \to \infty} \frac{\omega_{z,R}}{\omega_R} = 1,$$

uniformly in $z \in G$. Now it is obvious that

$$\operatorname{mes}(K_{z,R+R^\kappa} \setminus K_{z,R}) \leqslant \operatorname{mes}(V_{R+R^\kappa} \setminus V_R) = C_2 R^{n+\kappa-1}.$$

Hence we have

$$\frac{1}{\omega_R} \int \psi_R(x+z)\psi_R(x) f(x)dx = \frac{\omega_{z,R}}{\omega_R} \cdot \frac{1}{\omega_{z,R}} \int_{K_{z,R}} f(x)dx +$$

$$+ \frac{1}{\omega_R} \int_{K_{z,R+R^\kappa} \setminus K_{z,R}} \psi_R(x+z)\psi_R(x) f(x)dx.$$

Here the first summand on the right-hand side tends to $\mathbf{M}\{f\}$ uniformly in z, and the second is equal to $O(R^{\kappa-1})$. This implies (2.6).

Now we pass to the limit in (2.3) as $R \to \infty$. By Lemma 4.1 of [91] the limit of the left-hand side of (2.3) is equal to $\| T_\varepsilon u \|^2_{B^2}$. Hence, by (2.4)-(2.6) we obtain the inequality

$$\| T_\epsilon u \|_{B^2} \leqslant C \cdot \sup_{|z| \leqslant \epsilon} \| u(\cdot + z) - u(z) \|_{B^2}. \tag{2.7}$$

Taking into account that the group $\mathbf{R}^n$ acts continuously on $\mathbf{R}^n_B$ by translation, the corresponding shift operators $f(\cdot) \to f(\cdot + z)$ are strongly continuous with respect to $z \in \mathbf{R}^n$ in the space $B^2(\mathbf{R}^n) = L^2(\mathbf{R}^n_B, d\mu)$. This and (2.7) imply b). □

Now we are able to prove the following theorem, on the coincidence of weak and strong extensions.

THEOREM 2.3. *Let* $a(x, \xi) = a_1(x, \xi) + a_0(x, \xi)$, *where* $a_i \in \mathrm{APS}^i$, $i = 0, 1$. *Assume that* $a_1(x, \xi)$ *is positive homogeneous in* ξ *of order 1. Then* $\mathrm{CAP}^\infty(\mathbf{R}^n)$ *is dense in* $D(A)$ *with respect to the graph norm.*

Proof. Since $a_0(x, D)$ is bounded in $B^2(\mathbf{R}^n)$, we may assume that $a(x, \xi) = a_1(x, \xi)$. By Euler's theorem on homogeneous functions,

$$a(x, \xi) = \sum_{j=1}^n \xi_j \frac{\partial a}{\partial \xi_j}(x, \xi), \ \xi \neq 0.$$

Therefore,

$$a(x, D) = \sum_{j=1}^n b_j(x, D) \frac{\partial}{\partial x_j} \operatorname{mod} \mathrm{APL}^0.$$

where the $b_j(x, \xi)$ are positive homogeneous in ξ of order zero. Now Lemmas 2.1 and 2.2 imply that $\Phi_\epsilon u \to u$ with respect to the graph norm, and $\Phi_\epsilon u \in H^\infty(\mathbf{R}^n_B)$. Hence $H^\infty(\mathbf{R}^n_B)$ is dense in $D(A)$. But $\mathrm{CAP}^\infty(\mathbf{R}^n)$ is dense in $H^\infty(\mathbf{R}^n_B)$. □

Our main result is invertibility for first order operators:

THEOREM 2.4. *Under the assumptions of Theorem* 2.3 *the operator* $A = a(x, D)$ *is invertible in* $B^2(\mathbf{R}^n)$ *if and only if it is invertible in* $L^2(\mathbf{R}^n)$. *Moreover, the norms of* A^{-1} *in these spaces are equal.*

Proof. The formally adjoint operator $A^+ = a^+(x, D)$ also satisfies the assumption of the theorem. Hence A and A^+ are adjoint in $B^2(\mathbf{R}^n)$, regarded as unbounded operators. Here

$$D(A^+) = \{u \in B^2(\mathbf{R}^n) \,|\, A^+ u \in B^2(\mathbf{R}^n)\}.$$

Recall (see [1]) that $C_0^\infty(\mathbf{R}^n)$ is dense in the space

$$D(A, L^2) = \{v \in L^2(\mathbf{R}^n) \,|\, Av \in L^2(\mathbf{R}^n)\}.$$

The same is true for A^+. Hence A and A^+ are adjoint in $L^2(\mathbf{R}^n)$. Just as in [91, § 4] we have

$$\inf_{v \in C_0^\infty(\mathbf{R}^n)\setminus\{0\}} \frac{||Av||_{L^2}}{||v||_{L^2}} = \inf_{u \in \mathrm{CAP}^\infty(\mathbf{R}^n)\setminus\{0\}} \frac{||Au||_{B^2}}{||u||_{B^2}}, \tag{2.8}$$

and a similar inequality is valid for the adjoint operator. By Theorem 2.3 we may replace in (2.8) $C_0^\infty(\mathbf{R}^n)$ by $D(A;L^2)$ and $\mathrm{CAP}^\infty(\mathbf{R}^n)$ by $D(A)$. This implies the required statement, because the norms of A^{-1} coincide with the corresponding infima in (2.8). □

Theorem 2.4 allows us to transfer L^2-statements on invertibility to the a.p. situation. The simplest result of this kind is the following:

COROLLARY 2.5. *Let*

$$a(x, D) = \sum_{j=1}^{n} L_j(x)\frac{\partial}{\partial x_j} + M(x),$$

where $L_j^* = L_j$, $j = 1, \ldots, n$, *belong to* $\mathrm{CAP}^\infty(\mathbf{R}^n)$, *and let*

$$M(x) + M^*(x) - \sum_{j=1}^{n} \frac{\partial L_j(x)}{\partial x_j} \geqslant I,\ \alpha > 0.$$

Then $a(x, D)$ *is invertible in* $B^2(\mathbf{R}^n)$.

2.2. CAUCHY PROBLEM.

Consider the Cauchy problem

$$\mathfrak{L}u \equiv \frac{\partial u}{\partial t} - a(t, x, D)u = f(t, x), \tag{2.9}$$

$$u(0, x) = u_0(x). \tag{2.10}$$

The symbol $a(t, x, \xi)$ is assumed to be a continuously differentiable function of $t \in [0, T]$ with values in APS^{-1} (in the natural topology of the space of symbols). Let E be one of the spaces $L^2(\mathbf{R}^n)$ or $B^2(\mathbf{R}^n)$. Then with (2.9), (2.10) there are associated closed operators $\mathfrak{L}$ and $\mathfrak{L}_0$ acting in $L^2(0, T, E)$. The domains of these operators are, respectively,

$$D(\mathfrak{L}, E) = \{u \in L^2(0, T; E) \mid \mathfrak{L}u \in L^2(0, T; E)\}$$

and

$$D(\mathfrak{L}_0, E) = \{u \in D(\mathfrak{L}; E) \mid u(0) = 0\}.$$

The last definition makes sense. Indeed, if $u \in D(\mathfrak{L}; E)$, then $a(t, x, D)u \in L^2(0, T; E)$, where E_1 is equal to $H^{-1}(\mathbf{R}^n)$ or $H^{-1}(\mathbf{R}^n_B)$. Since $\mathfrak{L}u \in L^2(0, T; E) \subset L^2(0, T; E_1)$ we have $\partial u / \partial t \in L^2(0, T; E_1)$ and, by the trace theorem [46], $u \in C([0, T]; E_1)$. Therefore the equality $u(0) = 0$ makes sense.

Now we assume that

$$a(x, t, \xi) = a_1(t, x, \xi) + a_0(t, x, \xi),$$

where the a_i, $i = 0, 1$, are continuously differentiable functions with values in APS^i and a_1 is positively homogeneous in ξ of order 1. It is known that $C_0^\infty((0, T) \times \mathbf{R}^n)$ is dense in $D(\mathfrak{L}_0; L^2)$ with respect to the graph norm. We state here an a.p. version of this fact.

LEMMA 2.6. *The space*

$$\{v \in C^\infty([0, T]; \mathrm{CAP}^\infty(\mathbf{R}^n)) \mid v(0, x) = 0\}$$

is dense in $D(\mathfrak{L}_0; B)$ *with respect to the graph norm.*

Proof. The operators Φ_ϵ commute with $\partial / \partial t$ and all estimates in the proofs of Lemmas 2.1 and 2.2 may be obtained uniformly in $t \in [0, T]$. Hence the Φ_ϵ are uniformly bounded in $L^2(0, T; B^2(\mathbf{R}^n))$ and strongly converge to the identity. Moreover the commutators

$$[\mathfrak{L}, \Phi_\epsilon] = [a(t, x, D), \Phi_\epsilon]$$

are uniformly bounded in $L^2(0, T; B^2(\mathbf{R}^n))$ and strongly converge to zero. This implies in a standard manner that the $\Phi_\epsilon u$ converge to u with respect to the graph norm, provided $u \in D(\mathfrak{L}_0; B^2)$. For such u it is clear that for any $m \in \mathbf{Z}_+$,

$$\Phi_\epsilon u \in L^2(0, T; H^m(\mathbf{R}^n_B)).$$

Hence

$$\frac{\partial}{\partial t} \Phi_\epsilon u = \mathfrak{L}\Phi_\epsilon u - a(t, x, D)\Phi_\epsilon u \in L^2(0, T, B^2(\mathbf{R}^n)),$$

and $(\Phi_\epsilon u)(0)=0$. Using Bochner-Fejér operators with respect to $x\in\mathbf{R}^n$, we may approximate $\Phi_\epsilon u$ in the topology of the space

$$\{v\,|\,v\in L^2(0,T;H^m(\mathbf{R}^n_B))\,\forall m\in\mathbf{Z}_+,\ \frac{\partial v}{\partial t}\in L^2(0,T;B^2(\mathbf{R}^n)),\ v(0)=0\}$$

by a function belonging to the space

$$\{v\,|\,v\in L^2(0,T;\mathrm{CAP}^m(\mathbf{R}^n))\,\forall m\in\mathbf{Z}_+,$$

$$\frac{\partial v}{\partial t}\in L^2(0,T;B^2(\mathbf{R}^n)),\ v(0)=0\} \tag{2.11}$$

Hence this last space is dense in $D(fsL_0;B)$.

Now let $\theta_\epsilon\in C_0^\infty(\mathbf{R})$ be a nonnegative function such that

$$\int\theta_\epsilon(t)dt=1$$

and $\operatorname{supp}\theta_\epsilon\subset\{t\,|\,\epsilon\leqslant t\leqslant 2\epsilon\}$. Then for a member v of (2.11) we have

$$\theta_\epsilon\star v\in C^\infty([0,T];\mathrm{CAP}^\infty(\mathbf{R}^n))$$

and $(\theta_\epsilon\star v)(0)=0$. Moreover, it is not difficult to see that $\theta_\epsilon\star v\to v$ in the space (2.11) and, as a consequence, with respect to the graph norm of $D(\mathfrak{L}_0;B^2)$. □

Now we are able to obtain an analogue of Theorem 2.4.

THEOREM 2.7. *Under the assumptions stated above, the Cauchy problem* (2.9), (2.10) *is uniquely solvable in* $L^2(0,T;B^2(\mathbf{R}^n))$ *for any* $\{u_0,f\}\in B^2(\mathbf{R}^n)\times L^2(0,T;B^2(\mathbf{R}^n))$ *if and only if it is uniquely solvable in* $L^2((0,T)\times\mathbf{R}^n)$ *for any* $\{u_0,f\}\in L^2(\mathbf{R}^n)\times L^2((0,T)\times\mathbf{R}^n)$.

Proof. First we consider the case when $u_0=0$. Taking into account Lemma 2.6 and its version for $\mathfrak{L}_0^*$, we may proceed just as in the proof of Theorem 2.4. Therefore we see that $\mathfrak{L}_0^{-1}$ exists and is bounded in $L^2(0,T;B^2(\mathbf{R}^n))$ if and only if it exists and is bounded in $L^2(0,T;L^2(\mathbf{R}^n))=L^2((0,T)\times\mathbf{R}^n)$.

Now we return to the general case. Unique solvability of (2.9), (2.10) in $L^2(0,T;E)$ for any $\{u_0,f\}\in E\times L^2(0,T;E)$ implies the estimate

$$||u||_{L^2(0,T;E)}\leqslant C(||fsLu||_{L^2(0,T;E)}+||u||_E). \tag{2.12}$$

By Lemmas 4.1-4.3 of [91] and Lemma 2.6, the estimate (2.12) with $E=B^2(\mathbf{R}^n)$ is equivalent to the same estimate with $E=L^2(\mathbf{R}^n)$. Hence, unique solvability of the Cauchy problem with arbitrary data in the case $E=L^2(\mathbf{R}^n)$ (for the sake of being specific) implies that the operator

$$W: L^2(0, T; B^2(\mathbf{R}^n)) \to B^2(\mathbf{R}^n) \times L^2(0, T; B^2(\mathbf{R}^n)),$$

$$Wu = \{u(0), \mathfrak{L}u\},$$

has zero kernel and closed range. Now let $\{h, v\} \in B^2(\mathbf{R}^n) \times L^2(0, T; B^2(\mathbf{R}^n))$ be a vector orthogonal to $\operatorname{Im} W$. Then

$$\int_0^t (\mathfrak{L}u, v)_B dt = 0 \forall u \in C^\infty([0, T]; \mathrm{CAP}^\infty(\mathbf{R}^n)), \ u(0, x) = 0.$$

Hence $\mathfrak{L}_0^* v = 0$. Since $\mathfrak{L}_0$ and, as consequence, $\mathfrak{L}_0^*$ are invertible in $L^2(0, T; B^2(\mathbf{R}^n))$, we find that $v=0$. Now, by the orthogonality,

$$(u(0), h)_B = 0 \forall u \in C^\infty([0, T]; \mathrm{CAP}^\infty(\mathbf{R}^n)).$$

The traces at zero of such u's are clearly dense in $B^2(\mathbf{R}^n)$ and we have $h=0$. Therefore W is surjective. By changing the roles of $B^2(\mathbf{R}^n)$ and $L^2(\mathbf{R}^n)$ in the arguments above the proof is complete. □

Theorem 2.7 offers (as in n°. 2.2) the possibility to transfer all results on L^2-solvability of hyperbolic Cauchy problems to the B^2-setting. We describe here a most simple result of this kind (for the L^2-theory see [47]).

COROLLARY 2.8. *Let*

$$a(t, x, D) = \sum_{j=1}^n L_j(t, x) \frac{\partial}{\partial x_j} + M(t, x).$$

Suppose $L_j(t, x), j=1, \ldots, n, M(t, x)$ *are of class* $C^1([0, T], \mathrm{CAP}^\infty(\mathbf{R}^n))$ *and the* $L_j = L_j^*, j=1, \ldots, n,$ *are real-valued, or suppose* $a(t, x, D)$ *is strictly hyperbolic* [47]. *Then for any* $\{u_0, f\} \in B^2(\mathbf{R}^n) \times L^2(0, T; B^2(\mathbf{R}^n))$ *the problem* (2.9), (2.10) *has a unique solution* $u \in L^2(0, T; B^2(\mathbf{R}^n))$.

There is a similar assertion for Friedrichs- or Lax-symmetrizable systems [1, 129, 130].

REMARK 2.9. In case $a(t, x, \xi)=a(x, \xi)$ Theorem 2.7 follows immediately from Theorem 2.4 and the Hille-Yosida theorem. Indeed, by Theorem 2.4 the norms of the operators $(A^k-\lambda I)^{-1}$ in the spaces $B^2(\mathbf{R}^n)$ and $L^2(\mathbf{R}^n)$ are equal for any $k\in\mathbf{Z}_+$.

Theorem 2.7 may be applied to the Cauchy problem for a scalar equation

$$\mathfrak{L}u = \frac{\partial^m u}{\partial t^m}+\sum_{\substack{|\alpha|+j\leq m\\ j<m}} a_{\alpha j}(t, x)D_x^\alpha\frac{\partial^j u}{\partial t^j} = f, \tag{2.13}$$

$$u(0) = u_0,\ \frac{\partial u}{\partial t}(0) = u_1, \ldots, \frac{\partial^{m-1}u}{\partial t^{m-1}}(0) = u_{m-1}, \tag{2.14}$$

where $a_{\alpha j}\in C^1([0, T]; \mathrm{CAP}^\infty(\mathbf{R}^n))$.

Using a well-known method of Calderon [47, 52, 125] we may reduce this problem to the Cauchy problem for a first order pseudodifferential system. To do this we rewrite the leading part of $\mathfrak{L}$ in the form

$$\mathfrak{L}_m = \frac{\partial^m}{\partial t^m}+\sum_{j=1}^m h_j(t, x, D)\frac{\partial^{m-j}}{\partial t^{m-j}},$$

where

$$h_j(t, x, D) = \sum_{|\alpha|=j} a_{\alpha,m-j}(t, x)D^\alpha.$$

Denote by $\lambda_1(t, x, \xi), \ldots, \lambda_m(t, x, \xi)$ the roots of the characteristic polynomial

$$p(\lambda) = \lambda^m+\sum_{j=1}^m h_j(t, x, \xi)\lambda^{m-j}.$$

Let $H_j(t)$ be a pseudodifferential operator with symbol $h_j(t, x, \xi/|\xi|)$ $(j=1, \ldots, m)$. It is clear that $H_j(t)\in\mathrm{APL}^0$. Introduce new unknown functions by

$$v_j = [i(|D|+1)]^{m-j-1}\frac{\partial^j u}{\partial t^j},\ j = 0, 1, \ldots, m-1.$$

It is clear that

$$\frac{\partial v_j}{\partial t} = i(|D|+1)v_{j+1}.$$

We have

$$\mathcal{L}_m u = \frac{\partial}{\partial t}v_{m-1} + i\sum_{j=0}^{m-1} H_{m-j}(t)|D|(i|D|)^{m-j-1}\frac{\partial^j u}{\partial t^j}.$$

Let S_j be an operator with symbol

$$\left[\frac{|\xi|}{1+|\xi|}\right]^j - 1,\ j = 0, \ldots, m-1.$$

It is not difficult to see that $S_j \in \mathrm{APL}^{-1}$ and

$$(i|D|)^j = (1+S_j)[i(|D|+1)]^j = [i(|D|+1)]^j(1+S_j).$$

Then

$$\mathcal{L}_m u = \frac{\partial}{\partial t}v_{m-1} + i\sum_{j=0}^{m-1} H_{m-j}(t)|D|v_j + i\sum_{j=0}^{m-1} H_{m-j}(t)(|D|S_{m-1-j})v_j.$$

Hence (2.13) is equivalent to the system

$$\frac{\partial v}{\partial t} - i\mathcal{H}(t)|D|v + \mathcal{B}(t)v = \tilde{f}(t), \tag{2.15}$$

where $\tilde{f}(t) = (0, 0, \ldots, f(t))$, $v = (v_0, \ldots, v_{m-1})$,

$$\mathcal{H}(t) = \begin{bmatrix} 0 & 1 & 0 & \cdots & 0 & 0 \\ 0 & 0 & 1 & \cdots & 0 & 0 \\ \cdot & \cdot & \cdot & & \cdot & \cdot \\ \cdot & \cdot & \cdot & & \cdot & \cdot \\ \cdot & \cdot & \cdot & & \cdot & \cdot \\ 0 & 0 & 0 & \cdots & 0 & 1 \\ -H_m & -H_{m-1} & -H_{m-2} & \cdots & -H_2 & -H_1 \end{bmatrix},$$

and $\mathcal{B}(t) \in \mathrm{APL}^0$, since $|D|S_j \in \mathrm{APL}^0$. We may apply Theorem 2.7 to the system (2.15) (since all symbols involved are classical). In order to formulate the result we introduce the following notations:

$$\tilde{u}(t) = (u(t), \frac{\partial u}{\partial t}(t), \ldots, \frac{\partial^{m-1} u}{\partial t}(t)),$$

$$\tilde{H}(\mathbf{R}^n) = H^{m-1}(\mathbf{R}^n) \times \cdots \times H^0(\mathbf{R}^n),$$

$$\tilde{H}(\mathbf{R}^n_B) = H^{m-1}(\mathbf{R}^n_B) \times \cdots \times H^0(\mathbf{R}^n_B).$$

Then we have

THEOREM 2.10 *Problem* (2.13), (2.14) *has a unique solution* $\tilde{u} \in L^2(0, T; \tilde{H}(\mathbf{R}^n_B))$ *for any data* $\{\tilde{u}_0, f\} \in \tilde{H}(\mathbf{R}^n_B) \times L^2(0, T; B^2(\mathbf{R}^n))$ *iff it has a unique solution in the space* $L^2(0, T, \tilde{H}(\mathbf{R}^n))$ *for any* $\{\tilde{u}_0, f\} \in \tilde{H}(\mathbf{R}^n) \times L^2((0, T) \times \mathbf{R}^n)$.

Recall that $\mathfrak{L}$ is said to be strictly hyperbolic if all $\lambda_j(t, x, \xi)$ are real and

$$\inf_{\substack{(t,x) \in [0,T] \times \mathbf{R}^n \\ |\xi| = 1, j \neq k}} | \lambda_j(t, x, \xi) - \lambda_k(t, x, \xi) | > 0$$

Theorem 2.10 and classical L^2-results (see [47, Ch. 6) imply:

COROLLARY 2.11. *If* $\mathfrak{L}$ *is a strictly hyperbolic operator, then for any* $\{\tilde{u}_0, f\} \in \tilde{H}(\mathbf{R}^n_B) \times L^2(0, T;\ B^2(\mathbf{R}^n))$ *the Cauchy problem* (2.13), (2.14) *has a unique solution* $\tilde{u} \in L^2(0, T; \tilde{H}(\mathbf{R}^n_B))$.

Other results on a.p. strictly equations can be found in [29-31].

REMARK 2.12. All results of this section are still valid if we replace the spaces $L^2(\mathbf{R}^n) = H^0(\mathbf{R}^n)$ and $B^2(\mathbf{R}^n) = H^0(\mathbf{R}^n_B)$ by $H^s(\mathbf{R}^n)$ and $H^s(\mathbf{R}^n_B)$, respectively, where $s \in \mathbf{R}$ is arbitrary. Indeed, the invertibility of $a(x, D)$ in $H^S(\mathbf{R}^n)$ (respectively, in $H^S(\mathbf{R}^n_B)$) is equivalent to the invertibility of $(1 + | D |)^S \circ a(x, D) \circ (1 + | D |)^{-S}$ in the space $L^2(\mathbf{R}^n)$ (respectively, in $B^2(\mathbf{R}^n)$). The same is valid for the Cauchy problem as well.

REMARK 2.13. In fact, we may work with CAP^3-coefficients instead of CAP^∞ (even CAP^2 in the case of differential operators). For this certain modifications in the proofs of Lemmas 2.1, 2.2 and 2.6 are necessary. We must proceed as in n°. 2.1 of Ch. 4 and need not use the theorem on coincidence of norms.

3. Symmetric hyperbolic systems with monotone nonlinearity.

In this section we will present extensions of some of the results of § 2 to the case of nonlinear symmetric hyperbolic systems. In contrast to the case in the previous section, here we will consider real-valued functions and operators.

3.1. STATIONARY PROBLEMS.

Let $L_j(x), j=1, \ldots, n$, be real $(N\times N)$-matrices with $\mathrm{CAP}^\infty(\mathbf{R}^n)$- entries, and let $M(x)$ be an $(N\times N)$-matrix with $\mathrm{CAP}(\mathbf{R}^n)$-entries. The differential operator

$$L = \sum_{j=1}^{n} L_j(x)\frac{\partial}{\partial x_j}+M(x) \tag{3.1}$$

may be regarded as a closed operator from $B^p(\mathbf{R}^n)$ into $B^{p'}(\mathbf{R}^n)$ with domain

$$D_p(L) = \{u\in B^p(\mathbf{R}^n)\,|\,Lu\in B^{p'}(\mathbf{R}^n)\}. \tag{3.2}$$

Here the action of L is considered by means of the a.p. Sobolev spaces introduced in § 6 of Ch. 1. This operator L is the weak extension of (3.1) defined initially on $\mathrm{CAP}^\infty(\mathbf{R}^n)$. First of all we prove that in this situation weak and strong extensions coincide. To this end we slightly extend the results of § 2 on Friedrichs averaging. As in § 2, let $\phi\in C_0^\infty(\mathbf{R}^n)$ be an even nonnegative function such that $\phi(x)=0$ for $|x|\geqslant 1$ and

$$\int\phi(x)dx = 1.$$

We set $\phi_\epsilon(x)=\epsilon^{-n}\phi(x/\epsilon)$ and define the operators Φ_ϵ by the formula

$$(\Phi_\epsilon u)(x) = (\phi_\epsilon * u)(x) = \int\phi_\epsilon(x-y)u(y)dy = \int\phi_\epsilon(y)u(x-y)dy. \tag{3.3}$$

These operators are well-defined on $\mathrm{CAP}(\mathbf{R}^n)$.

LEMMA 3.1. $\Phi_\epsilon(\mathrm{CAP}^\infty(\mathbf{R}^n))\subset\mathrm{CAP}^\infty(\mathbf{R}^n)$ *and*

$$||\Phi_\epsilon u||_{M,p,B}\leqslant C_{m,p,\epsilon}||u||_{B^p},\ u\in\mathrm{CAP}^\infty(\mathbf{R}^n), \tag{3.4}$$

where $m\in\mathbf{Z}_+$, $1<p<\infty$ *and* $C_{0,p,\epsilon}=1$.

Recall that $||\cdot||_{m,p,B}$ is the norm in the space $W^{m,p}(\mathbf{R}^n_B)$.

Proof. By (3.3) it is easy to see that $\Phi_\epsilon(\mathrm{CAP}^\infty(\mathbf{R}^n))\subset\mathrm{CAP}^\infty(\mathbf{R}^n)$.

In the proof of (3.4) we restrict ourselves to the case $m=0$ (the general case is quite similar). By Hölder's inequality and Fubini's theorem we have

$$(2T)^{-n}\int_{K_T} dx \,|\int \phi_\epsilon(y)u(x-y)dy\,|^p \leqslant$$

$$\leqslant (2T)^{-n}\int_{K_T} dx(\int \phi_\epsilon(y)dy)^{p/p'}(\int \phi_\epsilon(y)\,|\,u(x-y)\,|^p dy) =$$

$$= \int dy\phi_\epsilon(y)(2T)^{-n}\int_{K_T} |\,u(x-y)\,|^p dx,$$

where $K_T=\{x\,|\,|\,x_j\,|\leqslant T, j=1,\ldots,n\}$. Using the fact that

$$\mathbf{M}\{f\} = \lim_{T\to\infty}(2T)^{-n}\int_{K_T} f(x,y)dx,\; f\in \mathrm{CAP}(\mathbf{R}^n),$$

uniformly in $y\in\mathbf{R}^n$, we obtain (3.4) for $m=0$. □

Thus Φ_ϵ can be extended to a uniformly bounded family of operators

$$\Phi_\epsilon: B^p(\mathbf{R}^n) \to B^p(\mathbf{R}^n).$$

Moreover, it is easy to see that $\Phi_\epsilon(B^p(\mathbf{R}^n))\subset W^{\infty,p}(\mathbf{R}^n_B)$.

LEMMA 3.2. *For* $u\in B^p(\mathbf{R}^n)$, $1<p<\infty$, *we have*

$$\lim_{\epsilon\to 0} \|\,\Phi_\epsilon u-u\,\|_{B^p} = 0. \tag{3.5}$$

Proof. The operators Φ_ϵ are uniformly bounded in $B^p(\mathbf{R}^n)$. Since $\lim\Phi_\epsilon u=u$ in $\mathrm{CAP}(\mathbf{R}^n)$ for any $u\in\mathrm{CAP}(\mathbf{R}^n)$ and this space is dense in $B^p(\mathbf{R}^n)$, we obtain (3.5). □

THEOREM 3.3. *The space* $\mathrm{CAP}^\infty(\mathbf{R}^n)$ *is dense in* $D_p(L)$ *with respect to the graph norm, provided* $p\geqslant 2$.

Proof. Multiplication by $M(x)$ is a bounded operator on $B^2(\mathbf{R}^n)$ and, since $p\geqslant 2$, from $B^p(\mathbf{R}^n)$ into $B^{p'}(\mathbf{R}^n)$. Hence, without loss of generality we may assume that $M=0$.

Now let $u \in D_p(L)$. Then for any $v \in \mathrm{CAP}^\infty(\mathbf{R}^n)$ we have the identity

$$(\Phi_\epsilon Lu, v)_B = (u, L^+ \Phi_\epsilon v)_B = (u, \Phi_\epsilon L^+ v)_B - (u, T_\epsilon^* v)_B, \tag{3.6}$$

where L^+ is the formally adjoint operator of L and $T_\epsilon^* = -[L^+, \Phi_\epsilon]$. The left-hand side of (3.6) is a continuous linear functional with respect to $v \in B^p(\mathbf{R}^n)$. By Lemma 2.2, T_ϵ^* is a bounded linear operator on $B^2(\mathbf{R}^n)$ which coincides with the adjoint operator of $T_\epsilon = [L^+, \phi_\epsilon]$. Therefore, by (3.6), $v \to (u, \Phi_\epsilon L^+ v)_B$ is a continuous linear functional on $B^p(\mathbf{R}^n)$. This implies that $\Phi_\epsilon u \in D_p(L)$ and

$$L\Phi_\epsilon u = \Phi_\epsilon Lu + T_\epsilon u, \; u \in D_p(L). \tag{3.7}$$

By Lemma 2.2, $\lim_{\epsilon \to 0} T_\epsilon u = 0$ in $B^2(\mathbf{R}^n) \subset B^{p'}(\mathbf{R}^n)$. Now (3.7) and Lemma 3.2 imply that

$$\lim_{\epsilon \to 0} \Phi_\epsilon u = u$$

with respect to the graph norm. Since, by Lemma 3.1, $\Phi_\epsilon(u) \in W^{\infty,p}(\mathbf{R}^n_B)$, we find that the last space is dense in $D_p(L)$. To conclude it remains to observe that $\mathrm{CAP}^\infty(\mathbf{R}^n)$ is dense in $W^{\infty,p}(\mathbf{R}^n_B)$ in the natural topology, which is stronger than the graph topology of $D_p(L)$. □

We say that an operator L is nonnegative, $L \geqslant 0$, if $(L\phi, \phi) \geqslant 0$ for any $\phi \in C_0^\infty(\mathbf{R}^n)$. Here $(\cdot, \cdot)$ is the usual inner product in $L^2(\mathbf{R}^n)$. In the case of symmetric operators ($L_j^* = L_j$, $j = 1, \ldots, n$) there is a natural nonnegativity condition:

$$M(x) + M^*(x) - \sum_{j=1}^{n} \frac{\partial L_i}{\partial x_j}(x) \geqslant 0. \tag{3.8}$$

Theorem 3.3. and Lemmas 4.1-4.3 in [91] imply the following assertions.

COROLLARY 3.4. *If $L \geqslant 0$, then for any $p \geqslant 2$ the inequality*

$$(Lu, u)_B \geqslant 0, \; u \in D_p(L), \tag{3.9}$$

is valid. Conversely, if (3.9) holds for some $p \geqslant 2$, then $L \geqslant 0$.

COROLLARY 3.5. *The adjoint of $L: B^p(\mathbf{R}^n) \supset D_p(L) \to B^{p'}(\mathbf{R}^n)$, $p \geqslant 2$, coincides with the operator L^+ defined on $D_p(L^+)$.*

COROLLARY 3.6. *If $L \geqslant 0$, then*

$$L:B^p(\mathbf{R}^n)\supset D_p(L) \to B^{p'}(\mathbf{R}^n),\ p \geqslant 2,$$

is a maximal monotone operator.

Proof. Since $L\geqslant 0$, then together with (3.9) the inequality

$$(L^* v, v)_B \geqslant 0,\ v\in D_p(L^+) = D(L^*),$$

is valid, implying the assertion required. □

Now let $A:\mathbf{R}^n\times\mathbf{R}^N\to\mathbf{R}^N$ be a continuous map such that for any compactum $K\subset\mathbf{R}^N$ the function $A(x,\eta)$ is a.p. in the first variable uniformly with respect to $\eta\in K$. Assume that A is monotone, i.e.

$$(A(x,\eta)-A(x,\lambda))\cdot(\eta-\lambda) \geqslant 0,\ \eta,\lambda\in\mathbf{R}^N, \tag{3.10}$$

and that the inequalities

$$|A(x,\eta)| \leqslant c_1|\eta|^{p-1}+c_2,\ \eta,\lambda\in\mathbf{R}^N, \tag{3.11}$$

$$A(x,\eta)\cdot\eta \geqslant \gamma_1|\eta|^p+\gamma_2,\ \eta\in\mathbf{R}^N, \tag{3.12}$$

are valid with constants $c_1>0, \gamma_1>0, c_2, \gamma_2\in\mathbf{R}$ not depending on $x\in\mathbf{R}$. (The dot denotes the usual inner product in $\mathbf{R}^N$.)

THEOREM 3.7. *Assume that the inequalities* (3.10)-(3.12) *are valid with* $p\geqslant 2$ *and* $L\geqslant 0$. *Then for any* $f\in B^{p'}(\mathbf{R}^n)$ *there is a solution* $u\in B^p(\mathbf{R}^n)$ *of the equation*

$$Lu+A(x,u) = f. \tag{3.13}$$

Proof. The operator

$$(\mathbf{A}v)(x) = A(x, v(x))$$

satisfies all the assumptions of § 2, Ch. 2. Hence $\mathbf{A}:B^p(\mathbf{R}^n)\to B^{p'}(\mathbf{R}^n)$ is a continuous bounded monotone operator satisfying the coerciveness inequality

$$(\mathbf{A}v, v)_B \geqslant \gamma_1\|v\|^p_{B^p}+\gamma_2.$$

By Corollary 3.6, L is a maximal monotone operator and Theorem 1.1 [44, Ch. 3] finishes the proof. □

REMARK 3.8. The uniqueness assertion holds in this case if

$$(a(x,\eta)-A(x,\lambda))\cdot(\eta-\lambda) \geqslant c(\eta,\lambda) \geqslant 0, \tag{3.14}$$

where $c(\eta, \lambda)$ does not depend on x and $c(\eta, \lambda) \neq 0$ for $\eta \neq \lambda$. In this case **A** is a strictly monotone operator.

REMARK 3.9. It is still not clear whether the previous result is valid if $1<p<2$. It is true if the coefficients L_j, $j=1, \ldots, n$, and M do not depend on $x \in \mathbf{R}^n$. This follows from the fact that the operators Φ_ϵ commute with L.

3.2. SPACIALLY A.P. CAUCHY PROBLEM.

Let $L_j(t,x) \in C^1([0,T]; \mathrm{CAP}^\infty(\mathbf{R}^n))$, $j=1, \ldots, n$, and $M(t,x) \in C([0,T], \mathrm{CAP}(\mathbf{R}^n))$ be real $(N \times N)$-matrices, and let

$$A:[0,T] \times \mathbf{R}^n \times \mathbf{R}^N \to \mathbf{R}^N$$

be a continuous map such that for any compactum $K \subset \mathbf{R}^N$ the function $A(t,x,\eta)$ is almost periodic in $x \in \mathbf{R}^n$ uniformly with respect to $(t,\eta) \in [0,T] \times K$.

Consider the equation

$$\mathfrak{L}u + A(t,x,u)' = f(t,x), \tag{3.15}$$

where

$$\mathfrak{L} = \frac{\partial}{\partial t} + L(t) = \frac{\partial}{\partial t} + \sum_{j=1}^{n} L_j(t,x)\frac{\partial}{\partial x_j} + M(t,x). \tag{3.16}$$

Equation (3.15) is understood in the sense of distributions on $(0, T)$ with values in $W^{-1,r}(\mathbf{R}^n_B)$ for suitable r.

Define a closed operator

$$\mathfrak{L}_0 : L^p(0,T; B^p(\mathbf{R}^n)) \to L^{p'}(0,t; B^{p'}(\mathbf{R}^n))$$

by

$$D(\mathfrak{L}_0) = D_p(\mathfrak{L}_0) = \{u \in L^p(0,T; B^p(\mathbf{R}^n)) \mid \mathfrak{L}u \in L^{p'}(0,T; B^{p'}(\mathbf{R}^n)),\ u(0)=0\},$$

$$\mathfrak{L}_0 u = \mathfrak{L}u,\ u \in D(\mathfrak{L}_0).$$

The requirement $u(0)=0$ here is meaningful. Indeed, assume that $u \in L^p(0, T; B^p(\mathbf{R}^n))$ and $\mathfrak{L}u \in L^{p'}(0, T; B^{p'}(\mathbf{R}^n))$. Since $L(\cdot)u(\cdot) \in L^p(0, T; W^{-1,p}(\mathbf{R}^n_B))$ we see that $\partial u / \partial t \in L^r(0, T; W^{-1,r}(\mathbf{R}^n_B))$, $r=\min(p, p')$. Hence $u \in C([0, T]; W^{-1,r}(\mathbf{R}^n_B))$ and $u(0)$ is well-defined.

LEMMA 3.10. *For $p \geqslant 2$ the space*

$$\{u \in C^\infty([0, T]: \mathrm{CAP}^\infty(\mathbf{R}^n)) \mid u(0, x) = 0\}$$

is dense in $D_p(\mathfrak{L}_0)$ with respect to the graph norm.

Proof. The proof is similar to the proof of Lemma 2.6 if Lemmas 3.1 and 3.2 are used instead of Lemma 2.1. □

THEOREM 3.11. *Suppose that $L(t)+\lambda_0 I \geqslant 0$, $t \in [0, T]$, for some $\lambda_0 \in \mathbf{R}$, and that the function $A(t, x, \eta)$ satisfies inequalities* (3.10)-(3.12) *with $p \geqslant 2$ (uniformly with respect to $(t, x) \in [0, T] \times \mathbf{R}^n$). Then for any $u_0 \in B^2(\mathbf{R}^n)$ and any $f \in L^{p'}(0, T; B^{p'}(\mathbf{R}^n))$ there is a solution $u \in L^p(0, T; B^p(\mathbf{R}^n)) \bigcap C([0, T; B^2(\mathbf{R}^n))$ of* (3.15) *such that $u(0)=u_0$.*

Proof. Using the substitution $u=e^{\lambda_0 t}v$ we may assume that $L(t) \geqslant 0$. Moreover, the new $A(t, x, \eta)$ still satisfies all conditions of the theorem.

By Lemma 3.10, the adjoint of $\mathfrak{L}_0$ is defined by

$$D_p(\mathfrak{L}_0^*) = \{v \in L^p(0, T; B^p(\mathbf{R}^n)) \mid \mathfrak{L}^+ v \in L^{p'}(0, T; B^{p'}(\mathbf{R}^n)), v(T) = 0\}$$

$$\mathfrak{L}_0^* v = \mathfrak{L}^+ v, \ v \in D_p(\mathfrak{L}_0^*).$$

Moreover, the space

$$\{v \in C^\infty([0, T; \mathrm{CAP}^\infty(\mathbf{R}^n)) \mid v(T) = 0\}$$

is dense in $D_p(\mathfrak{L}_0^*)$ with respect to the graph norm.

We will show that $\mathfrak{L}_0$ is a maximal monotone operator. To this end we must prove the inequality

$$\int_0^T (\mathfrak{L}_0 u, u)_B dt \geqslant 0, \ u \in D_p(\mathfrak{L}_0), \tag{3.18}$$

and the similar inequality for $\mathfrak{L}_0^*$, By Lemma 3.10 it suffices to verify inequality (3.18) on the space (3.17). This may be done by a simple computation using the condition $L(t)\geqslant 0$ and Corollary 3.4. The estimate for the adjoint can be obtained in a similar manner.

The existence proof can be reduced to the case $u_0=0$. Indeed, if $u_0\in W^{1,p}(\mathbf{R}^n_B)$, then reduction to the case of homogeneous initial data is evident. Since $W^{1,p}(\mathbf{R}^n_B)$ is dense in $B^2(\mathbf{R}^n)$, the general case may be considered by passage to the limit, using inequality (3.19) below.

In case $u_0=0$ the Cauchy problem for (3.15) is equivalent to the equation

$$\mathfrak{L}_0 u+\mathbf{A}u = f,$$

and the existence of a solution follows from Theorem 1.1 [44, Ch. 3].

For $\lambda_0=0$ any two solutions $u_i\in L^p(0, T; B^p(\mathbf{R}^n))\cap C([0, T]; B^2(\mathbf{R}^n))=E$, $i=1, 2$, of (3.15) satisfy the inequality

$$\| u_1(t)-u_2(t) \|_{B^2} \leqslant \| u_1(0)-u_2(0) \|_{B^2},\ t\in[0, T], \tag{3.19}$$

which implies uniqueness. To prove (3.19) we set $u_{i\epsilon}=\Phi_\epsilon u_i$ and

$$f_{i\epsilon} = \mathfrak{L}u_{i\epsilon}+\mathbf{A}u_{i\epsilon}.$$

Then $\lim u_{i\epsilon}=u_i$ in E and $\lim f_{i\epsilon}=f$ in $L^{p'}(0, T; B^{p'}(\mathbf{R}^n))$. A simple computation leads to the following identity

$$\int_0^t (f_{1\epsilon}-f_{2\epsilon}, u_{1\epsilon}-u_{2\epsilon})_B d\tau =$$

$$= \frac{1}{2}\| u_{1\epsilon}-u_{2\epsilon} \|^2_{B^2}\Big|_0^t+\int_0^t (L(u_{1\epsilon}-u_{2\epsilon})+\mathbf{A}u_{1\epsilon},\ u_{1\epsilon}-u_{2\epsilon})_B d\tau.$$

Corollary 3.4, the nonnegativity of L and the monotonicity of $\mathbf{A}$ imply that the integral on the right-hand side of this identity is nonnegative. Hence

$$\| u_{1\epsilon}-u_{2\epsilon} \|^2_{B^2}\Big|_0^t \leqslant 2\int_0^t (f_{1\epsilon}-f_{2\epsilon},\ u_{1\epsilon}-u_{2\epsilon})_B d\tau.$$

By passing to the limit we obtain (3.19). □

REMARK 3.12. The substitution $u=e^{(\lambda_0+\alpha)t}v$, $\alpha>0$, reduces (3.15) to the equation

$$\mathfrak{L}v+\alpha v+\tilde{A}(t,x,v) = \tilde{f}(t,x).$$

This permit us to consider the case slow growth of A in $\eta (p<2)$.

REMARK 3.13. If the coefficients of $L(t)$ and the nonlinearity $A(t,x,\eta)$ are a.p. in (t,x) we may apply Theorem 3.7. This implies the existence of a solution which is Besicovitch a.p. in (t,x). Moreover, choosing $H=B^2(\mathbf{R}^n)$, $V=B^2(\mathbf{R}^n)$ we may apply the results of § 1, Ch. 4. Thus, a solution of (3.15) can be found in the space $S^p(\mathbf{R};B^p(\mathbf{R}^n))\cap \mathrm{CAP}(\mathbf{R};B^2(\mathbf{R}^n))$, provided $f\in S^{p'}(\mathbf{R};B^{p'}(\mathbf{R}^n))$ and

$$(A(t,x,\eta)-A(t,x,\lambda))\cdot(\eta-\lambda) \geqslant \alpha|\eta-\lambda|^p,\ p\geqslant 2.$$

This solution is asymptotically stable in the space $B^2(\mathbf{R}^n)$ (cf. Proposition 2.10 of Ch. 3). In fact, for any two solutions $u_i\in BS^p(\mathbf{R}_+;B^p(\mathbf{R}^n))\cap C_b(\mathbf{R}_+,B^2(\mathbf{R}^n))\,(i=1,2)$ with such a right-hand side f we have $u_1-u_2\in L^p(\mathbf{R}_+;B^p(\mathbf{R}^n))$ and $\lim_{t\to+\infty}||u_1(t)-u_2(t)||_{B^2}=0$.

REMARK 3.14. The regularity conditions for the leading coefficients may be weakened to $\mathrm{CAP}^2(\mathbf{R}^n)$ (cf. Remark 2.13)

4. A nonlinear Schrödinger-type equation.

In this section we will consider the equation

$$\frac{\partial u}{\partial t}-i\Delta u+a(u) = f(t,x),\ (t,x)\in\mathbf{R}\times\mathbf{R}^n, \tag{4.1}$$

where f and u are complex-valued functions. Here $a:\mathbf{C}\to\mathbf{C}$ is a continuous map such that for any $z,w\in\mathbf{C}$,

$$(a(z)-a(w))\cdot(\bar{z}-\bar{w}) \geqslant 0, \tag{4.2}$$

$$|a(z)| \leqslant c_1|z|^{p-1}+c_2, \tag{4.3}$$

$$a(z)\cdot\bar{z} \geqslant \alpha|z|^p+\beta, \tag{4.4}$$

where $c_1>0$, $\alpha>0$, $c_2,\beta\in\mathbf{R}$, and $p>1$.

At first we discuss the spatially a.p. Cauchy problem for (4.1) with initial data

$$u(0,x) = u_0(x). \tag{4.5}$$

(Recall that time almost periodicity of solutions of the boundary value problem for equation (4.1) was studied in § 2 of Ch. 4.)

THEOREM 4.1. *Let $p \geq 2$. Then for any $u_0 \in B^2(\mathbf{R}^n)$ and any $f \in L^{p'}(0, t; B^{p'}(\mathbf{R}^n))$ there exists a unique solution $u \in L^p(0, T; B^p(\mathbf{R}^n)) \bigcap C([0, T]; B^2(\mathbf{R}^n))$ of the problem* (4.1), (4.5).

The *proof* is similar to that of Theorem 3.11. We need only verify the maximal monotonicity of the operator $\mathcal{L}_0$ defined by

$$\mathcal{L}_0 u = \frac{\partial v}{\partial t} - i\Delta v,$$

$$D(\mathcal{L}_0) = \{v \in L^p(0, T; B^p(\mathbf{R}^n)) \,|\, v' - i\Delta v \in L^{p'}(0, T; B^{p'}(\mathbf{R}^n)),\ v(0) = 0\}$$

It is not difficult to see that the memberships $v \in L^p(0, T; B^p(\mathbf{R}^n))$ and $v' - i\Delta v \in L^{p'}(0, T; B^{p'}(\mathbf{R}^n))$ imply that $v \in C([0, T]; B^2(\mathbf{R}^n))$. Thus $u(0)$ is well-defined.

Now we will consider the space

$$\{v(t,x) = \sum a_k(t) \exp i\xi_k x \,|\, a_k \in C^\infty([0, T],\ a_k(0) = 0\}$$

(each v containing only a finite number of summands). We prove that this space is dense in $D(\mathcal{L}_0)$ endowed with the graph norm. Since the Bochner-Fejér operators L_γ (with respect to the x-variables) commute with d/dt - $i\Delta$, we see that $\lim L_\gamma v = v$ with respect to the graph norm, for any $v \in D(\mathcal{L}_0)$. Hence $D(\mathcal{L}_0)$ contains a dense subset, consisting of the trigonometric polynomials

$$v = \sum a_k(t) \exp i\xi_k \cdot x \tag{4.6}$$

such that $a_k \in L^p(0, T)$, $a_k' \in L^{p'}(0, T)$ and $a_k(0) = 0$. Now we need only approximate such v by $\phi_\epsilon \star v$ (convolution with respect to t), where $\phi_\epsilon \in C_0^\infty(\mathbf{R})$ is a nonnegative function such that $\operatorname{supp} \phi_\epsilon \subset \{t \,|\, \epsilon \leq t \leq 2\epsilon\}$ $(\epsilon > 0)$ and

$$\int \phi_\epsilon(t) dt = 1.$$

Define the operator $\mathfrak{M}_0$ by

$$\mathfrak{M}_0 v = -\frac{\partial v}{\partial t} + i\Delta v,$$

$$D(\mathfrak{M}_0) = \{v \in L^p(0, T; B^p(\mathbf{R}^n)) \mid v' - i\Delta v \in L^{p'}(0, T; B^{p'}(\mathbf{R}^n)),\ v(T) = 0\}.$$

Just as for $\mathfrak{L}_0$, the functions (4.6) with $a_k \in C^\infty([0, T])$, $a_k(T) = 0$, form a dense subset of $D(\mathfrak{M}_0)$ (with respect to the graph norm). This implies that

$$(\mathfrak{L}_0 v, w) = (v, \mathfrak{M}_0 w),\quad v \in D(\mathfrak{L}_0),\quad w \in D(\mathfrak{M}_0),$$

i.e. $\mathfrak{L}_0^* = \mathfrak{M}_0$, and

$$(\mathfrak{L}_0 v, v) \geqslant 0,\quad v \in D(\mathfrak{L}_0),$$

$$(\mathfrak{M}_0 w, w) \geqslant 0,\quad w \in D(\mathfrak{M}_0).$$

Here

$$(u, v) = \int_0^T (u, v)_B\, dt.$$

It is sufficient to verify the last statement for trigonometric polynomials of the type considered above (instead of for $v \in D(\mathfrak{L}_0)$ and $w \in D(\mathfrak{M}_0)$), and this can be done by straightforward computation. □

REMARK 4.2. Assume that the estimates (4.2) and (4.3) with $p \leqslant 2$ are valid. The substitution $u = e^{\alpha t} v$ $(\alpha > 0)$ reduces (4.1) to

$$\frac{\partial v}{\partial t} - i\Delta v + \alpha v + e^{-\alpha t} a(e^{\alpha t} v) = e^{-\alpha t} f(t, x)$$

(cf. Remark 3.12). Then the nonlinearity

$$\tilde{a}(t, v) = \alpha v + e^{-\alpha t} a(e^{\alpha t} v)$$

satisfies inequalities (4.2)-(4.4) with $p = 2$ and Theorem 4.1 applies. Evidently here the dependence of $\tilde{a}$ on t is inessential.

Now we will consider solutions of (4.1) which are a.p. in all variables. The following statement is valid (here $1<p<\infty$).

THEOREM 4.3. *For any* $f\in B^{p'}(\mathbf{R}^n)$ *there is a solution* $u\in B^p(\mathbf{R}^n)$ *of equation* (4.1).

Proof. Just as in the proof of Theorem 3.7 it is sufficient to prove (keeping in mind to use Theorem 1.1 [44, Ch. 3]) that $\mathfrak{L}$, defined by

$$D(\mathfrak{L}) = \{v\in B^p(\mathbf{R}^{n+1})\,|\,v' - i\Delta v\in B^{p'}(\mathbf{R}^{n+1})\},$$

$$\mathfrak{L}v = \frac{\partial v}{\partial t} - i\Delta v,$$

is a maximal monotone operator. Actually, $\mathfrak{L}$ is skewadjoint, i.e. $\mathfrak{L}^* = -\mathfrak{L}$. Indeed, for $v, w\in \mathrm{Trig}(\mathbf{R}^{n+1})$ the equality

$$(\mathfrak{L}v, w)_B = -(v, \mathfrak{L}w)_B \tag{4.7}$$

may be verified by straightforward computation, and we must show that

$$D(\mathfrak{L}) = D(\mathfrak{L}^*). \tag{4.8}$$

By definition, $\mathfrak{L}$ is a weak extension of the operator

$$\frac{\partial}{\partial t} - i\Delta\colon \mathrm{Trig}(\mathbf{R}^{n+1}) \to B^{p'}(\mathbf{R}^{n+1}).$$

Hence $\mathfrak{L}^*$ is a strong extension of

$$-\frac{\partial}{\partial t} + i\Delta\colon \mathrm{Trig}(\mathbf{R}^{n+1}) \to B^{p'}(\mathbf{R}^{n+1}),$$

and $\mathrm{Trig}(\mathbf{R}^{n+1})$ is dense in $D(\mathfrak{L}^*)$ (with respect to the graph norms). Now, (4.8) follows from denseness of $\mathrm{Trig}(\mathbf{R}^{n+1})$ in $D(\mathfrak{L})$. This last statement may be obtained by using Bochner-Fejér operators (with respect to all variables), which clearly commute with $\mathfrak{L}$. □

REMARK 4.4. Now let $p\geq 2$. Assume that

$$(a(z) - a(w))\cdot(\bar{z} - \bar{w}) \geq \alpha\,|\,z - w\,|^p,\ z, w\in \mathbf{C}$$

(for example, $a(z) = |\,z\,|^{p-2}z$). In this case the results of § 1, Ch. 4, apply. Hence for any $f \in S^{p'}(\mathbf{R}; B^{p'}(\mathbf{R}^n))$ equation (4.1) has a unique solution $u \in S^p(\mathbf{R}; B^p(\mathbf{R}^n)) \bigcap \mathrm{CAP}(\mathbf{R}, B^2(\mathbf{R}^n))$. This solution is asymptotically stable in the space $B^2(\mathbf{R}^n)$. Moreover, the situation here is similar to that in Remark 3.13. Thus there is a striking contrast with the case of an analytic nonlinearity, for which a large number of periodic solutions exists [10]. The same also holds for the one-dimensional Schrödinger equation with nonlinearity $\pm i \mid z \mid^2 \cdot z$, which is integrable by the inverse scattering method [20].

Comments

The results of § 1 were obtained in [63, 69]. It is of interest that the maps $v \to \psi_R \cdot v$ determine discrete approximations [56] (see also [7]) of Sobolev- Besicovitch spaces by ordinary Sobolev spaces. In addition, the operator A_B is approximated by A. The results of [56] imply that if for A_B, and as consequence for A, the p-monotonicity and p-coerciveness condition are fulfilled, then a.p. solutions of $A_B u = f$ can be discretely approximated by ordinary solutions of $Au = \psi_R f$. Moreover, in the proof of Theorem 1.1 we have constructed, in fact, discrete approximations of ordinary Sobolev spaces by Sobolev-Besicovitch spaces and of the operator A by certain a.p. operators.

In § 2 the results of [61] are presented. They extend theorems on the coincidence of spectra in [90] to certain nonhypoelliptic cases. § 3 is based on the paper [64]. The results of § 4 were obtained in [65]. Other related results on nonlinear hyperbolic systems and nonlinear Schrödinger equations can be found in [104, 139, 143].

APPENDIX 1.

On certain linear evolution equations.

1. ALMOST PERIODICITY IN THE SENSE OF BESICOVITCH.

Let $H^1 \subset H^0$ be complex Hilbert spaces. We assume H^0 equals its antidual space, so $H^0 \subset H^{-1} = (H^1)'$. Here the symbol "$'$" stands for taking the antidual space, i.e. the space of continuous antilinear functionals. Let $A(t): H^1 \to H^{-1}$, $t \in \mathbf{R}$, be a family of bounded linear operators and let

$$a(t; u, v) = (A(t)u, v), \quad u, v \in H^1,$$

denote the corresponding semilinear form on H^1. We assume that the following conditions hold:

(i) the complex-valued function $t \mapsto a(t; u, v)$ is a.p. in sense of Bohr for any $u, v \in H^1$;

(ii) the inequality

$$\| A(t)u \|_* \leqslant c \| u \|, \quad u \in H^1, \tag{1}$$

is valid, where $c > 0$ does not depend on t and u;

(iii) the form $a(t; u, v)$ is coercive uniformly in t, i.e.

$$\mathrm{Re}\, a(t; u, v) \geqslant \alpha \| u \|^2, \quad u \in H^1, \ y \in \mathbf{R}, \tag{2}$$

where $\alpha > 0$ does not depend on t.

In particular, $A(t)$ is an a.p. operator function with respect to the weak operator topology. The theory of such operator functions may be developed along the same lines as the theory of weakly a.p. functions (see § 1 of Ch. 5). In our case, using the Hilbert space structure of H^1 and H^{-1}, it can be proved that the spectrum* of the a.p. function $A(t)$ is countable and $A(t)$ can be extended to an operator function on $\mathbf{R}_B$ which is continuous in the weak operator topology. A slight modification of

* Do not confuse this with the spectrum of an operator.

the arguments of § 2 of Ch. 2 shows that the operator $\mathbf{A}$ defined by $(\mathbf{A}u)(t)=A(t)u(t)$ is a bounded linear operator acting from $\mathcal{H}^1=B^2(\mathbf{R};H^1)$ into $\mathcal{H}^{-1}=B^2(\mathbf{R};H^{-1})$, and let

$$\mathrm{Re}(\mathbf{A}u, u)_B \geqslant \alpha \,||\, u \,||_B^2. \tag{3}$$

Also, we set $\mathcal{H}^0=B^2(\mathbf{R};H)$.

Now we will consider the equation

$$\frac{du}{dt}+A(t)u = f(t). \tag{4}$$

Here the derivative is meant in the sense of § 5 of Ch. 3.

THEOREM 1. *For any* $f\in B^2(\mathbf{R};H^{-1})$ *equation* (4) *has a unique solution* $u\in B^2(\mathbf{R};H^1)$.

The *proof* reduces to a straightforward verification of the conditions of the isomorphism theorem for a sum of operators [46, Ch. 3, Theorem 1.1]. Let $G(s)$ be the group of right shifts:

$$[G(s)f](t) = f(t-s)$$

acting on the spaces $\mathcal{H}^i$, $i=-1,0,1$. In all these spaces, $G(s)$ is a strongly continuous group of unitary operators. If $-\Lambda$ is its generator, we have $\Lambda=d/dt$ with domain defined by

$$D_i(\Lambda) = D(\Lambda;\mathcal{H}^i) = \{u\in\mathcal{H}^i \,|\, \frac{du}{dt}\in\mathcal{H}^i\},\ i = -1,0,1.$$

Since $G^*(s)=G(-s)$, we have $D_i(\Lambda^*)=D_i(\Lambda)$ and $\Lambda^*=-\Lambda$. Now, by (3) and the isomorphism theorem we see that $\Lambda+\mathbf{A}$ isomorphically maps $D_{-1}(\Lambda)\cap\mathcal{H}^1$ into $\mathcal{H}^{-1}$.

By the adjoint isomorphism theorem [46, Ch. 3], $\Lambda+\mathbf{A}$ isomorphically maps $\mathcal{H}^1$ onto $(\mathcal{H}^1\cap D_{-1}(\Lambda^*))'=(\mathcal{H}^1\cap D_{-1}(\Lambda))'$. Here Λ is extended to $\mathcal{H}^1$ by duality as follows. Since $\Lambda^*\in L(\mathcal{H}^1\cap D_{-1}(\Lambda^*),\mathcal{H}^{-1})$ we have $\Lambda^{**}\in L(\mathcal{H}^1,(\mathcal{H}^1\cap D_{-1}(\Lambda))'$ and $\Lambda^{**}u=\Lambda u$ for any $u\in\mathcal{H}^1\cap D_{-1}(\Lambda)$. Now it is not difficult to see that

$$(D_{-1}(\Lambda))' = \{v\in D'(\mathbf{R}_B;H^1)\,|\,v = v_1+\frac{dv_2}{dt},\ v_i\in\mathcal{H}^1,\ i = 1,2\}.$$

Hence

$$(\mathcal{H}^1 \cap D_{-1}(\Lambda))' = \mathcal{H}^{-1} + (D_{-1}(\Lambda))' =$$

$$= \{v \in \mathcal{D}'(\mathbf{R}_B; H^{-1}) \mid v = v_1 + \frac{dv_2}{dt},\ v_1 \in \mathcal{H}^{-1},\ v_2 \in \mathcal{H}^1\}$$

and we obtain

THEOREM 2. *Let $f = f_1 + df_2/dt$, where $f_1 \in B^2(\mathbf{R}; H^{-1})$ and $f_2 \in B^2(\mathbf{R}; H^1)$. Then equation* (4) *has a unique solution $u \in B^2(\mathbf{R}; H^1)$.*

The next step consists of interpolation. By a well-known complex interpolation theorem, $\Lambda + \mathbf{A}$ is an isomorphism between $[\mathcal{H}^1 \cap D_{-1}(\Lambda), \mathcal{H}^1]_\theta$ and $[\mathcal{H}^{-1}, (\mathcal{H}^1 \cap D_{-1}(\Lambda))']_\theta$, where $[X, Y]_\theta$ denotes the complex interpolation functor (see, for example, [3]). Now we will describe explicitly all spaces involved. For $-1 \leqslant \alpha \leqslant 1$ we set

$$H^\alpha = \begin{cases} [H^1, H^0]_{1-\alpha}, & 0 \leqslant \alpha \leqslant 1, \\ [H^0, H^{-1}]_{1+\alpha}, & -1 \leqslant \alpha \leqslant 0. \end{cases}$$

It is obvious that $(H^\alpha)' = H^{-\alpha}$. Let B be a positive definite selfadjoint operator on H^0 with domain H^1. Then B is diagonalizable. Hence H^0 is isometrically isomorphic to a direct integral

$$\int_\oplus H(\lambda) d\mu(\lambda),$$

where $\mu(\lambda)$ is a positive measure on $[\lambda_0, +\infty)$, $\lambda_0 > 0$, and $H(\lambda)$ is a μ-measurable field of Hilbert spaces. Moreover, the operator B turns into the operator of multiplication by λ (see, for example, [74]). Under this isomorphism the space H^α is identified with the space of $H(\lambda)$-valued measurable functions having finite norm

$$\left[\int_{\lambda_0}^{\infty} \lambda^{2\alpha} \| \phi(\lambda) \|^2_{H(\lambda)} d\mu(\lambda) \right]^{1/2}.$$

The Fourier-Bohr transform isomorphically maps $B^2(\mathbf{R}; H)$ onto $l^2(\mathbf{R}; H^0)$, and the image of $B^2(\mathbf{R}; H^\alpha)$ consists of the $\{\hat{\phi}_\xi(\lambda)\}$ such that $\{\lambda^\alpha \hat{\phi}_\xi(\lambda)\} \in l^2(\mathbf{R}; H)$.

Moreover, the Fourier-Bohr image of $D_{-1}(\Lambda)$ consists of the $\{\hat{\phi}_\xi(\lambda)\}$ such that

$$\{(1+|\xi|)\lambda^{-1}\hat{\phi}_\xi(\lambda)\}\in l^2(\mathbf{R};H^0).$$

Hence, the Fourier-Bohr image of the space $[D_{-1}(\Lambda),\mathcal{H}^1]_\theta$ can be characterized as the space of those $\{\hat{\phi}_\xi(\lambda)\}$ for which

$$\{(1+|\xi|)^{1-\theta}\lambda^{2\theta-1}\hat{\phi}_\xi(\lambda)\}\in l^2(\mathbf{R};H^0).$$

This is equivalent to:

$$\{(1+|\xi|)^{1-\theta}\hat{\phi}_\xi\}\in l^2(\mathbf{R};H^{2\theta-1}).$$

Now, for $-1\leqslant\alpha,\beta\leqslant 1$ we define the space

$$H^\alpha(\mathbf{R};H^\beta) = \{\phi\in\mathcal{D}'(\mathbf{R}_B;H^{-1})\,|\,\{(1+|\xi|)^\alpha\hat{\phi}_\xi\}\in l^2(\mathbf{R};H^\beta),$$

and endow it with the norm

$$||\phi||_{\alpha,\beta} = \left[\sum_\xi(1+|\xi|)^{2\alpha}||\hat{\phi}_\xi||_\beta^2\right]^{1/2}.$$

Then

$$[D_{-1}(\Lambda),\mathcal{H}^1]_{1-\alpha} = H^\alpha(\mathbf{R};H^{1-2\alpha}),$$

$$[D_{-1}(\Lambda),\mathcal{H}^1]'_\alpha = H^{\alpha-1}(\mathbf{R};H^{1-2\alpha}).$$

Hence

$$[\mathcal{H}^1\cap D_{-1}(\Lambda),\mathcal{H}^1]_{1-\alpha} = B^2(\mathbf{R};H^1)\cap H^\alpha(\mathbf{R};H^{1-2\alpha}),$$

$$[\mathcal{H}^{-1},(fsH^1\cap D(\Lambda))']_{1-\alpha} = B^2(\mathbf{R};H^{-1})+H^{\alpha-1}(\mathbf{R};H^{1-2\alpha}).$$

This implies:

THEOREM 3. *For any* $f\in B^2(\mathbf{R};H)+H^{-s}(\mathbf{R};H^{2s-1})$, $0\leqslant s\leqslant 1$, *equation* (4) *has a unique solution* $u\in B^2(\mathbf{R};H^1)\cap H^{1-s}(\mathbf{R};H^{2s-1})$.

These results were obtained in [55].

2. Impulse evolution equations.

In a separable Banach space E we consider the equation

$$\frac{du}{dt} = A(t)u + f(t), \tag{5}$$

where $A(t)$ is a closed operator depending almost periodically on t, and $f(t)$ is an a.p. impulse measure of the form

$$f(t) = \sum_{k \in \mathbf{Z}} s_k \delta_{t_k}, \; t_k \in \mathbf{R}, \; s_k \in E. \tag{6}$$

Here δ_a denotes the Dirac δ-measure supported at the point $a \in \mathbf{R}$. It is assumed that $D(A(t)) \equiv D$ is a Banach space, densely and continuously embedded into E and that $A(t)$ is an a.p. function with values in the Banach space $L(D, E)$.

To formulate subsequent assumptions we consider the corresponding homogeneous equation

$$\frac{du}{dt} = A(t)u. \tag{7}$$

We assume that the Cauchy problem for (7) is uniformly well- posed on any finite interval [36] and that the corresponding evolution operator $U(t, \tau)$, $t \geqslant \tau$, satisfies the estimate

$$||\, U(t, \tau) \,||_{L(E)} \leqslant c\exp(a \,|\, t - \tau \,|), \; t \geqslant \tau. \tag{8}$$

Moreover, we also assume that $U(t, \tau) \in L(D)$ and that an estimate similar to (8) is valid with $L(E)$ replaced by $L(D)$. Similar assumptions, with the Cauchy problem replaced by the inverse Cauchy problem, are assumed to hold for the equation

$$\frac{dv}{dt} = -A^*(t)v, \tag{9}$$

which is adjoint to (7). We note that these assumptions are still valid for a perturbed equation

$$\frac{du}{dt} = A(t)u + B(t)u,$$

where $B(t)$ is an a.p. operator function with values in $L(E) \cap L(D)$.

Before we will deal with equation (5), we will briefly discuss certain discrete equations. Consider the homogeneous equation

$$u_{n+1} = T_n u_n,\ n \in \mathbf{Z}. \tag{11}$$

Here it is assumed that $\{T_n\}$ and $\{T_n^*\}$ are strongly a.p. sequences of operators, i.e. strongly a.p. functions on $\mathbf{Z}$ with values in the spaces $L(E)$ and $L(E')$, respectively. The evolution operator $X_{n,k}$, $n \geqslant k$, $n, k \in \mathbf{Z}$, of (11) is defined by

$$X_{n,k} = \begin{cases} I, & n-k, \\ T_{n-k} \circ T_{n-2} \circ \cdots \circ T_k, & n > k. \end{cases} \tag{12}$$

For solutions of the corresponding inhomogeneous equation

$$u_{n+1} = T_n u_n + s_n,\ n \in \mathbf{Z},\ s_n \in E, \tag{13}$$

the formula

$$u_n = X_{n,n_0} u_{n_0} + \sum_{n_0}^{n}{}_k X_{n,k} S_k,\ n \geqslant n_0, \tag{14}$$

Holds. Here, by definition,

$$\sum_{n_0}^{n}{}_k a_k = \begin{cases} \sum\limits_{k=n_0+1}^{n} a_k, & n > n_0, \\ 0, & n = n_0, \\ -\sum\limits_{k=n+1}^{n_0} a_k, & n < n_0. \end{cases}$$

The evolution operator $X_{n,k}$ may be extended to the domain $\{n<k\}$ by the formula

$$X_{n,k} = T_n^{-1} \circ T_{n+1}^{-1} \circ \cdots \circ T_{k-1}^{-1},\ n < k, \tag{15}$$

provided all operators T_n, $n \in \mathbf{Z}$, are invertible. The representation (15) still holds for $n < n_0$.

Together with equations (11) and (12) it is useful to consider the equations

$$u_{n+1} = T_n^{(m)} u_n, \; n \in \mathbf{Z}, \; m \in \mathbf{Z}_B, \tag{16}$$

where $T_n^{(m)} = \tilde{T}_{m+n}$, and $\{\tilde{T}_m\}_{m \in \mathbf{Z}_B}$ is a continuous extension of $\{T_n\}$ to $\mathbf{Z}_B$.

PROPOSITION 1. *Assume that all nontrivial bounded solutions of equations* (16) *are semiseparated from zero, i.e.* $\inf_{n \leqslant 0} \| u_n \| > 0$ *for any such solution* $\{u_n\}$. *Let* $\{s_n\}$ *be a weakly a.p. sequence. Then if equation* (13) *has a bounded solution, it also has a weakly a.p. solution.*

PROPOSITION 2. *Assume that for any* $n \in \mathbf{Z}$, $T_n^{-1} \in L(E)$ *exists, the sequences* $\{T_n\}$ *and* $\{T_n^{-1}\}$ *are a.p. with respect to the operator norm in* $L(E)$, *there is a constant* $C > 0$ *such that*

$$\| X_{n,k} \| < C, \; n \geqslant k, \tag{17}$$

and that $\{s_n\}$ *is an a.p. E-valued sequence. Then if (13) has a bounded solution, it also has an a.p. solution.*

These statements are discrete versions of results of Favard type (compare with [25, 41]) and their proofs are similar to those for differential equations. In particular, weakly a.p. solutions may be constructed by the well-known minimax method. More precisely, a bounded solution with minimal $\sup_{n \in \mathbf{Z}} \| u_n \|$ is weakly a.p.

Now we return to impulse equations. We assume that the sequence $\{t_k\} \subset \mathbf{R}$ (see (6)) is strictly increasing and that $\lim_{k \to \pm\infty} t_k = \pm\infty$. The measure $f(t)$ defined by (6) is said to be (weakly) a.p. if $\{s_k\}$ is a (weakly) a.p. E-valued sequence and if all sequences $\{t_k^{(j)}\}$ defined by $t_k^{(j)} = t_{k+j} - t_k$ are a.p. uniformly in $j \in \mathbf{Z}$. A solution of (5) is a function $u(t)$ of the form

$$u(t) = U(t, t_k) u_k, \; y_k \leqslant t \leqslant t_{k+1}, \tag{18}$$

such that

$$u_{k+1} = U(t_{k+1}, t_k) u_k + s_{k+1} \tag{19}$$

for any $k \in \mathbf{Z}$. A function $u(t)$ with jumps at points $\{t_k\}$ is said to be a.p. if for any $\epsilon > 0$ there is a relatively dense subset $\{\eta\} \in \mathbf{R}$ such that

$$\| u(t+\eta) - u(t) \| \leqslant \epsilon, \; | t - t_k | \geqslant \epsilon, \; k \in \mathbf{Z}.$$

Such a function is said to be weakly a.p. if for any $v \in E'$ the real-valued function $(v, u(t))$ is a.p. in the above mentioned sense. Any function which is (weakly) a.p. in that sense is (weakly) a.p. in the sense of Stepanov (see [83a]).

We will also consider equations

$$\frac{dv}{dt} = A^{(l)}(t)v, \; l \in \mathbf{R}_B, \tag{20}$$

where $A^{(l)}(t) = \bar{A}(l+t)$ and $\bar{A}$ is a continuous extension of A to $\mathbf{R}_B$.

THEOREM 3. *Assume that all nontrivial bounded solutions of equations* (20) *are semiseparated from zero, i.e.* $\inf_{t<0} \| v(t) \| > 0$ *for any such solution* v. *Let* $f(t)$ *be a weakly a.p. impulse measure. Then if equation* (5) *has a bounded solution, it also has a weakly a.p. solution.*

The *proof* is carried out along the following lines. First it may be proved that the evolution operator $U(t, \epsilon)$ is strongly a.p. on the diagonal. This means that the operator function $V(h) = U(t+h, \tau+h)$ is a.p. in the strong operator topology uniformly with respect to pairs (t, τ) for which $t-\tau \in J$, where J is a finite interval. The same is also true for $U^*(t, \tau)$. These facts imply that the sequences $\{T_k\}$ and $\{T_k^*\}$, where $T_k = U(t_{k+1}, t_k)$, are strongly a.p. Now it may be proved that equations (16) with such $\{T_k\}$ satisfy the semiseparatedness condition. Moreover, the existence of bounded solution of (5) implies the same for equation (13). Now Proposition 1 permits us to construct a weakly a.p. sequence $\{u_k\}$ satisfying equation (19). The function $u(t)$ defined by (18) is a solution of (5), and it may be proved that it is weakly a.p. (technically this is the most complicated part of the proof). □

Similarly, Proposition 2 permits us to obtain the following:

THEOREM 4. *Assume that* $D = E$, $f(t)$ *is a.p. and there is a constant* $C > 0$ *such that*

$$\| U(t, \epsilon) \| \leqslant C, \; t \geqslant \tau. \tag{21}$$

Then if (5) has a bounded solution, it also has an a.p. solution.

The condition $D = E$ is very restrictive. It excludes any application to partial differential equations. It is not clear whether it is possible to relax the last condition and to obtain a natural version of the corresponding continuous result ([41, Ch. VIII, § 5]). It may be that to this aim almost periodicity must be regarded in the sense of Stepanov and not as in the above.

THEOREM 5. *Assume that the embedding* $D \subset E$ *is compact, that* $U(t+h, \tau+h)$ *is a.p. in* $L(E)$ *as a function of* h *uniformly with respect to* (t, τ), $t-\tau \in [0, T]$ *for ant*

$T>0$, and that estimate (21) *holds. Assume also that $U(t,\tau)\in L(E,D)$, $t>\tau$ and*

$$||\,U(t,\tau)\,||_{L(E,D)} \leqslant \phi(t-\tau), \tag{22}$$

where $\phi(\eta)$ is a continuous function on $\eta>0$. Let $f(t)$ be a.p. and $t_n^{-1}\geqslant\theta>0$ for any $n\in\mathbf{Z}$. Then, if equation (5) *has a bounded solution, it also has an a.p. solution.*

Condition (22) holds for abstract parabolic equations, at least in the case $A(t)\equiv A$. The same is true with respect to the condition of almost periodicity of the evolution operator mentioned in Theorem 5. For parabolic partial differential equations in a bounded spatial domain the compactness condition is valid too.

A.p. solutions of discrete and impulse equations in finite-dimensional spaces were studied in papers of Halanay and Wexler (see, for example, [83a]). They stated that such a solution exists and is unique, provided the trivial solution of the corresponding homogeneous equation is uniformly asymptotically stable. It can be shown that this result is still valid for exponentially dichotomic equations. Proofs of the results presented here are given in [59].

APPENDIX 2.

On certain wave equations.

First we will consider an abstract hyperbolic equation with "nonlinear friction",

$$u'' + \phi(u') + Au = f, \tag{1}$$

in a real Hilbert space H. We assume that $A = Q^2$, where Q is a positive definite selfadjoint linear operator with compact inverse and that ϕ is a monotone non-linear operator with $\phi(0) = 0$. Denote by H_1 the domain $D(Q)$ of Q endowed with norm $|u|_1 = |Qu|$. Suppose also that we are given a reflexive Banach space E such that $E \subset H \subset E'$. We assume that

a) $\phi: E \to E'$ is a continuous monotone operator;

b) $H_1 \subset E$ continuously.

In the energy space $V = H_1 \times H$ with norm $||(u, v)||^2 = |u_1|^2 + |v|^2$, equation (1) may be written as

$$z' + Lz = g, \tag{2}$$

where $z = (u, v)$, $Lz = (-v, Au + \phi(v))$ with $D(L) = \{z \mid u \in D(A), v \in D(Q)\}$, and $g = (0, f)$. It can be shown that for any locally Lipschitz V-valued function $g(t)$ equation (2) has a unique classical solution $z(t)$ with arbitrarily given initial value $z(t_0) \in D(L)$. Moreover, for such solutions $z_1(t), z_2(t)$ with initial values $z_1(t_0), z_2(t_0)$ and right-hand sides $g_1(t), g_2(t)$ the estimate

$$||z_1(t) - z_2(t)|| \leqslant ||z_1(t_0) - z_2(t_0)|| + \int_{t_0}^{t} ||g_1 - g_2|| ds \tag{3}$$

is valid. This estimate permits us to obtain a generalized solution for any $z(t_0) \in V$ and any $g \in L_{loc}^{-1}([t_0, \infty))$ by closing the set of all classical solutions.

Now assume that

$$\lim_{||u||_E \to \infty} \frac{(\phi u, u)}{||u||_E} = +\infty, \tag{4}$$

$$||\phi u||_E^\alpha \leqslant k_1(\phi u, u)\cdot\gamma(|u|)+k_2, \tag{5}$$

where $\alpha\geqslant 1$ and $\gamma(s)$ is nondecreasing function such that $\lim_{s\to+\infty}\gamma(s)/s=0$.

The following assertion provides a condition for boundedness of solutions.

PROPOSITION 1. *Let* $f\in C_b(\mathbf{R}_+;H)$ *and let conditions* a), b), (4) *and* (5) *with* $\alpha=1$ *be satisfied. Then for any solution u we have*

$$||u(t)||^2 \leqslant \max\{||u||^2_{L^\infty(0,2,v)}, C\},$$

where C does not depend on u.

Now we assume that $f\in \mathrm{CAP}(\mathbf{R};H)$ and state some results on almost periodic solutions.

THEOREM 2. *Let the embedding* $E\subset H$ *be compact and conditions a), b), (4) and (5) with* $\alpha>1$ *be satisfied. Then equation* (1) *has a solution* $u\in\mathrm{CAP}(\mathbf{R};V)$.

THEOREM 3. *Assume that conditions* a), b), *and the inequalities*

$$(\phi(u+v), u) \geqslant \gamma_1(||v||_E),$$

$$||\phi(u+v)-\phi(u)||_{E'} \leqslant k\gamma_1(||v||_E)\gamma(|v|)+k_2$$

are satisfied, where $\gamma(s)$ *and* $\gamma_1(s)$ *are nondecreasing functions such that* $\gamma_1(s)>0$ *for* $s>0$ *and* $\gamma(s)/s\to 0$, $\gamma_1(s)/s\to 0$ *as* $s\to\infty$. *Then equation (1) has a unique solution* $u\in\mathrm{CAP}(\mathbf{R};V)$.

The proofs can be found in [25].

Now we will consider the wave equation

$$\frac{\partial^2 u}{\partial t^2}+\phi\left(\frac{\partial u}{\partial t}\right)+Au = f(t,x),\ (t,x)\in\mathbf{R}\times\Omega, \tag{6}$$

where A is a uniformly elliptic (formally) selfadjoint second order operator in a bounded domain $\Omega\times\mathbf{R}^n$ with regular boundary, i.e.

$$Au = -\sum_{i,j=1}^{n}\frac{\partial}{\partial x_j}\left(a_{ij}(x)\frac{\partial u}{\partial x_j}\right)+a_0(x)u,$$

$a_{ij}(x)=a_{ji}(x)$ are real-valued functions of class $C(\overline{\Omega})$, and $a_0\in L^\infty(\Omega)$, $a_0(x)\geqslant 0$. On the boundary the Dirichlet condition

$$u|_{\partial\Omega} = 0 \tag{7}$$

is posed.

For problem (6), (7) some results of conditional nature (if there is a precompact solution, then there is an a.p. solution), even for discontinuous nonlinearity ϕ, are contained in [100, 111]. Under more restrictive assumptions there are certain "unconditional" results, which are described below.

Let $\phi \in C(\mathbf{R})$ be a nondecreasing function such that $\phi(0)=0$. Assume that for some $p \geqslant 2$ the inequality

$$c_1|\xi|^p - c_0 \leqslant \xi\cdot\phi(\xi) \leqslant c_2|\xi|^p + c_0 \tag{8}$$

is valid with $c_i > 0 (i=0,1,2)$. The following result on solvability of the Cauchy problem was stated by Lions and Strauss [44]. Let $u_0 \in H_0^1(\Omega)$, $u_1 \in L^2(\Omega)$ and $f \in L^{p'}(0,T;L^{p'}(\Omega))$. Then the problem (6), (7) has a unique (weak) solution $u \in L^\infty(0,T;H_0^1(\Omega))$ such that $u' \in L^\infty(0,T;L^2(\Omega)) \cap L^p(0,T;L^p(\Omega))$, $u(0)=u_0$, and $u'(0)=u_1$. As above, it is useful to introduce the energy space $V=H_0^1(\Omega)\times L^2(\Omega)$ and to consider this solution as a couple $z(t)=(u(t),u'(t))$.

Now we assume that

$$2 \leqslant p \leqslant 2 + \frac{4}{n-1}, \tag{9}$$

$$c_3|\xi_1-\xi_2|^{p-2} \leqslant \frac{\phi(\xi_1)-\phi(\xi_2)}{\xi_1-\xi_2} \leqslant c_4(1+|\xi_1|^{p-2}+|\xi_2|^{p-2}). \tag{10}$$

Then we have

THEOREM 4. *Let $f \in \mathrm{BS}^{p'}(\mathbf{R};L^{p'}(\Omega)$. Then there is a unique solution u of (6), (7) such that $z=(u,u') \in C_b(\mathbf{R};V)$, $u' \in B^p(\mathbf{R};L^p(\Omega)$. Moreover, if $f \in S^{p'}(\mathbf{R};L^{p'}(\Omega))$, then $z \in \mathrm{CAP}(\mathbf{R};V)$ and $u' s^p(\mathbf{R};L^p(\Omega))$.*

We note also that under the assumptions of Theorem 4 the solution is asymptotically stable in the energy sense, i.e. if $\bar{u}$ is an another solution on $\mathbf{R}_+$, then $\lim_{t\to\infty}||z(t)-\bar{z}(t)||=0$, where $z=(\bar{u},\bar{u}')$. Proofs of these (and other) results can be found in [100].

A typical example of a nonlinearity is $\phi(\xi)=c|\xi|^{p-2}\cdot\xi$. From the point of view of applications (a nonlinear membrane) the case $\phi(\xi)=\gamma_1\xi+\gamma_2|\xi|\xi (\gamma_1>0, \gamma_2>0)$ is of interest. If $2 \leqslant n \leqslant 5$, then for such ϕ conditions (9), (10) are satisfied and Theorem 4 covers the physical case $n=2$.

Finally we note that much more detailed information on a.p. solutions of nonlinear wave equations is contained in [131a].

APPENDIX 3.

Open questions.

Here we formulate some questions whose answers are of interest in the theories of a.p. functions and a.p. differential equations.

1. Are Proposition 5.7 and Theorem 5.14 of Ch. 1 still valid if we do not assume the space E to be weakly sequentially complete or reflexive, respectively?

2. It is of interest to systematically study interpolations of spaces of bounded and Stepanov a.p. functions. It can be shown that

$$[BS^{p_0}, BS^{p_1}]^\theta = BS^p(\mathbf{R}),$$

$$[S^{p_0}(\mathbf{R}), S^{p_1}(\mathbf{R})]_\theta = S^p(\mathbf{R}),$$

where $1/p=(1-\theta)/p_0+\theta/p_1, 0<\theta<1$. Here $[\cdot,\cdot]_\theta$ and $[\cdot,\cdot]^\theta$ stand for the first and the second complex interpolation method, respectively [3]. What are the spaces $(BS^{p_0}(\mathbf{R}), BS^{p_1}(\mathbf{R}))_{\theta,r}$ and $(S^{p_0}(\mathbf{R}), S^{p_1}(\mathbf{R}))_{\theta,r}$ constructed by the real method? This question is closely related to the following one.

3. The spaces $BS^p(\mathbf{R})$ can be regarded as a special case of the following general construction. Let E be a Banach space of measurable functions on (0, 1). We denote by BSE the space of all measurable functions on $\mathbf{R}$ having finite norm

$$||f|| = \sup_{t\in\mathbf{R}} ||f(t+\cdot)||_E.$$

If E contains the nonzero constant, then BSE $\neq \{0\}$ (it also contains the nonzero constants). Conversely, if E is an ideal Banach lattice and BSE $\neq \{0\}$, then E contains the constant functions (in fact, $L^\infty(\mathbf{R})\subset$BSE). In this setting the space SE may be defined as the closure of Trig($\mathbf{R}$) in BSE. As far as is known to the author, a systematic treatment of such spaces is lacking in the literature. In particular, the interpolation of spaces BSE and SE is of interest. It is also of interest to investigate the properties of BSE and SE as Banach spaces and Banach lattices. It is known that they are nonreflexive and nonseparable, but BSE contains, evidently, l^∞, while for SE this is not true (at least for $E=L^p(0, 1)$). Are they weakly compactly generated (WCG) spaces (for a definition see [18])*?

Among the concrete spaces E, Lorentz spaces [3] (in connection with question 2) and Orlicz spaces [33] (in connection with applications) are most interesting to study. It is also useful to investigate the action of singular integral operators (in particular, Hilbert operators) on the spaces BSE and SE.

4. Let $\Lambda \subset \mathbf{R}$. Under what conditions on Λ is the subspace $S^p_\Lambda(\mathbf{R})$ closed in $B^p(\mathbf{R})$?

In other words, when there is an estimate

$$||f||_{S^p} \leqslant C||f||_{B^p},\ f \in \mathrm{Trig}_\Lambda(\mathbf{R}),$$

where C does not depend on f? (The opposite estimate is always valid.) Another formulation of this question is as follows. When is every Besicovitch a.p. function f with $\mathrm{sp}(f) \subset \Lambda$ a.p. in the sense of Stepanov?

There is the trivial example $\Lambda = \alpha\mathbf{Z}$ (periodic functions).*

5. Let $\Pi(a, b) = \{z \in \mathbf{C} \mid z = x + iy,\ a < y < b\}$. Recall that a holomorphic function f in $\pi(a, b)$ is said to be a.p. if for any α and $\beta (a < \alpha < \beta < b)$ and for any $\epsilon > 0$ the set of $\tau \in \mathbf{R}$ such that

$$\sup_{z \in \Pi(\alpha, \beta)} |f(z+\tau) - f(z)| \leqslant \epsilon$$

is relatively dense. Further, a meromorphic function in a plane domain may be viewed simply as a holomorphic map from that domain into the complex projective line $\mathbf{CP}^1$ (the Riemann sphere). Let ρ be the usual spherical metric on $\mathbf{CP}^1$. It is natural to define a meromorphic a.p. function on $\Pi(a, b)$ as a holomorphic map $f: \pi(a, b) \to \mathbf{CP}^1$ such that for any $\alpha, \beta \in (a, b)$, $\alpha < \beta$, and for any $\epsilon > 0$ the set of ϵ-almost periods, i.e. those $\tau \in \mathbf{R}$ which satisfy

$$\sup_{z \in \Pi(\alpha, \beta)} \rho(f(z+\tau), f(z)) \leqslant \epsilon,$$

is relatively dense. The investigation of meromorphic a.p. function is of interest. One concrete problem is the following. Is it true that any meromorphic a.p. function is the quotient of (two) holomorphic a.p. functions?

6. It is well-known [40], that for a.p. functions that are holomorphic in the half-plane $\{\mathrm{Im}\, z > 0\}$ there is a version of the classical Picard theorem. For such a

* A Plicko informs us that these spaces are not WCG (added in translation).

* In case $p = 2$ some results of this kind can be found in [40] (added in translation).

function f one of the following three statements is valid.

A. $\lim_{y\to\infty} f(x+iy)$ exists and is finite, uniformly in $x \in \mathbf{R}$;

B. $\lim_{y\to\infty} |f(x+iy)| = +\infty$, uniformly in $x \in \mathbf{R}$;

C. In any half-plane $\{\operatorname{Im} z > \alpha\}$ the function f take all its values except, possibly, one. Is there a qualitative refinement of this result by means of Nevanlinna theory [86]? Instead of the classical function $N(r, a)$ (the averaged number of a-points, i.e. solutions of $f(z)=a$, in the disc $\{|z| \leqslant r\}$) it is natural to consider the r-averaged density of the a-points in the strip $\Pi(\alpha, r)$. We also note that meromorphic a.p. functions are more natural in this setting that holomorphic a.p. functions.

7. The definition of almost periodicity can clearly be extended to holomorphic functions on tube subdomains of $\mathbf{C}^n$. Such a.p. functions have in fact never been studied (in connection with this see, however, [12]). For example, it is well-known that the envelope of holomorphy of a tube domain

$$T_B = \{x+iy \in \mathbf{C}^n \mid x \in \mathbf{R}^n, y \in B\},$$

where B is a domain in $\mathbf{R}^n$, is the domain $T_{\hat{B}}$, $\hat{B} = \operatorname{conv} B$ (convex hull). Any holomorphic a.p. function on T_B is can be extended to a holomorphic a.p. function on $T_{\hat{B}}$. Let B be a convex domain. What are natural a.p. analogues of the spaces $H^p(T_B)$ [81]? For example, let $HS^p(T_B)$ be the space of holomorphic a.p. functions in T_B for which

$$\sup \| f(\cdot + iy) \|_{S^p} < \infty.$$

What can be said about the boundary values of functions from this space? This question is closely related to the following problem. Which a.p. functions can appear as boundary values of functions harmonic in a half-space? The case of a half-plane was studied by Favard (see [40]). It is also of interest to study systems of conjugate harmonic (in the sense of [81]) a.p. functions. There is another related question. What singular integral operators act continuously in the space $BS^p(\mathbf{R}^n)$ and $S^p(\mathbf{R}^n)$? Are these operators the same as for $L^p(\mathbf{R}^n)$? (For L^p-theory see, for example, [80, 81].)

8. The problem of "almost periodicity without compactness" [25, 44] mentioned in Ch. 3 and Ch. 4 has not been solved in full generality. The following question arises. Can one eliminate conditions (2.6) and (2.7) from Theorems 2.5 and 3.9 of Ch. 3 (or, which is the same, eliminate compactness of the embedding $V \subset H$ from Theorems 4.3 and 4.8)?

Similar questions with respect to Theorems 1.9 and 1.12 of Ch. 4 are even more important from the point of view of applications. Moreover, it is useful to relax condition ($H5$) in Theorem 1.12 of Ch. 4.

It is of interest also to have an answer to the following more special question. Let u_λ be a bounded solution of the problem (4.5), Ch. 3. Under the conditions of § 4 of Ch. 3, do the u_λ converge to a solution of (1.2) in the space $C_b(\mathbf{R}; H)$?

9. The general theory presented in Ch. 3 and Ch. 4 deals with the case of coercive operators $A(t)$. Can one construct bounded a.p. solutions if that condition is relaxed somehow? Apparently, this question is very difficult in general. So it is of interest to investigate various special one-sided problems (see, for example, [21]) that are coercive in a weaker sense, and to state sufficient conditions for the existence of a.p. solutions. Some examples of this kind are considered in § 7 of Ch. 3 (see, also, [131a]).

10. There is an important unsolved problem concerning averaging in evolution variational inequalities. Consider an a.p. variational inequality of the form

$$(u'(t)+A(\epsilon^{-1}t)u(t)-f(\epsilon^{-1}t),\, v-u(t))-\phi(v)-\phi(u(t)) \geqslant 0,\ \forall v\in V,$$

or its weak version. Is it true that its a.p. solutions converge in some sense to a solution of the averaged inequality

$$(\bar{A}u_0-\bar{f},\, v-u_0)+\phi(v)-\phi(u_0) \geqslant 0,\ \forall v\in V,$$

as $\epsilon\to 0$? Here

$$Av = \mathbf{M}\{A(t)v\} = \lim_{T\to\infty}\frac{1}{2T}\int_{-T}^{T} A(t)v\,dt,$$

$$\bar{f} = \mathbf{M}\{f(t)\} = \lim_{T\to\infty}\frac{1}{2T}\int_{-T}^{T} f(t)\,dt.$$

The following difficulty arises. Usually, justification of the averaging procedure essentially requires some smoothness of the operators involved in the problems under consideration [24]. For ordinary differential equations with monotone operators bounded in one space, H, the smoothness condition was dropped in [71]. But variational inequalities are problems essentially involving nonsmooth operators acting between different spaces (from V into V'). It might be useful to regard a variational inequality as an equation with multivalued operator, using then nonlinear semigroup theory [118, 119, 151].

11. For the general nonlinear system (3.3), (3.4) of Ch. 4, can one prove the existence of bounded and Stepanov or Bohr a.p. solutions? The same question is of interest for problem (3.16), (3.17) as well

12. G. Prouse [100] proved the existence of a.p. solution of the problem

$$u'' - \Delta u - |u'|^{p-2} \cdot u' = f, \; u|_{\partial\Omega} = 0,$$

where $\Omega \subset \mathbf{R}^n$ and $f(t)$ is a.p., provided

$$2 \leqslant p \leqslant 2 + \frac{4}{n-1}.$$

Can one construct an a.p. solution (even in the sense of Besicovitch) without this restriction? We note here that for the corresponding variational inequalities [44, 118] a.p. solutions have not been studied at all.

13. Consider the wave equation

$$u'' - \Delta u + f(t, u) = 0,$$

with periodic or Dirichlet boundary conditions in spatial variables. It is of interest to find sufficient conditions for the existence of almost periodic or quasi-periodic (in time) nontrivial solutions, provided $f(t, 0) \equiv 0$. The case of a "small" nonlinearity $f(t, u) = \epsilon g(t, u)$ was considered by V.B. Moseenkov [48, 49]. For periodic solutions (in the case of one spatial variable) see, for example, [123, 146] and the references given there.

14. Do the norms (and spectra) of a.p. pseudodifferential operators in the spaces $L^p(\mathbf{R}^n)$ and $B^p(\mathbf{R}^n)$ coincide? The case $p = 2$ was studied originally in [90, 91] (see also Ch. 5). The probable answer in general is yes. Can this be proved using the technique of n°. 1.1, Ch. 5?

15. For a.p. nonlinear elliptic equations the monotonicity method gives rise to very weak a.p. solutions (in the sense of Sobolev-Besicovitch spaces). Can one obtain general results on the existence of classical bounded solutions with certain almost periodicity properties? For second order equations something of this kind was done in n°. 1.2 of Ch. 5, but the higher order case is of interest too. A more concrete question is the following. For a.p. equations of the form

$$Au = f(x, u, D_n, \ldots, D^{m-1}u),$$

where A is a linear elliptic operator of order m, in [77] a theorem on the existence of a classical bounded solution has been obtained. Can such a solution belong to the space $B^{\infty}(\mathbf{R}^n)$ as well?

16. For linear strictly hyperbolic equations the solvability of the a.p. Cauchy problem in the spaces of Stepanov and Bohr a.p. functions has been proved [30, 31]. Are there similar results for strictly hyperbolic and symmetric (or symmetrizable) hyperbolic systems? The same question is of interest also for hyperbolic systems with monotone nonlinearities.

17. For an equation of the form

$$\sum_{j=1}^{n} a^j \frac{\partial u}{\partial x_j} + g(x, u) = 0,$$

where a^j are constants and g is a function that is 2π-periodic in x and increasing in u, the following necessary and sufficient condition for the existence of a periodic solution was proved in [121]. Let P be the orthoprojection operator onto the kernel of the linear part of this equation (in L^2, on the torus). A periodic solution exists iff there are constants $\delta > 0$, m and M such that $Pg(x, M) \leqslant -\delta$, $Pg(x, m) \geqslant \delta$. Is there an extension of this statement to a.p. functions?

18. Nonlinear differential equations with homogeneous random coefficients are interesting, and unstudied, objects. All problems treated in the present text make sense for such random equations. For linear elliptic operators with random homogeneous coefficients see [13, 14, 16, 70]. Nonlinear random operators (not homogeneous) are discussed in [133, 134, 136, 137].

References

1. Agranovich, M.S.: Boundary value problems for systems of pseudo-differential operators, *Uspekhi Mat. Nauk* **24**, no. 1 (1969), 61-125 (in Russian).
2. Beckenbach, E.F. and Bellman, R.: *Inequalities*, Springer, 1961.
3. Bergh, J. and Löfström, J.: *Interpolation Spaces*, Springer, 1976.
4. Bourbaki, N.: *Intégration*, Chapters 6,7,8, Hermann, 1959-1963.
5. Bourbaki, N.: *Intégration*, Chapters 3,4,5,9, Hermann, 1965-1969.
6. Vainberg, M.M.: *Variational Method and Method of Monotone Operators in the Theory of Nonlinear Equations*, Wiley, 1973 (transl. from the Russian).
7. Vainikko, G.: *Analysis of Discretization Methods*, Tartusk Univ., 1976 (in Russian).
8. Vasil'eva, A.B. and Butuvov, V.F.: *Asymptotic Expansion of Singularly Perturbed Equations*, Nauka Moscow, 1973 (in Russian).
9. Vishik, M.I.: Quasilinear strongly elliptic systems of differential equations having divergent form, *Tr. Moskov. Mat. Obshch.* **12** (1963), 125-184 (in Russian).
10. Vishik, M.I. and Furshikov, A.V.: Cauchy's problem for a nonlinear equation of Schrödinger type, *Mat. Sb.* **96**, no. 4 (1975), 458-470 (in Russian).
11. Vishik, M.I. and Furshikov, A.V.: *Mathematical Problems of Statistical Hydrodynamics*, Nauka Moscow, 1980 (in Russian).
12. Gel'fond, O.A.: The mean indices of an almost periodic vector wave, *Preprint F.I. Akad. Nauk.* SSSR **219** (1982) (in Russian).
13. Gusev, A.I.: State density and other spectral invariants of selfadjoint elliptic operators with random coefficients, *Mat. Sb.* **104**, no. 2 (1977), 207- 226 (in Russian).
14. Gusev, A.I. and Shubin, M.A.: Elliptic operators with random coefficients. In: *Limit Theorems for Stochastic Processes*, Nauk. Dumka Kiev, 1977, 98-107 (in Russian).
15. Dunford, N., Schwartz, J.T.: *Linear Operators. General Theory*, Interscience, 1958.
16. Dedik, P.E. and Shubin, M.A.: Random pseudodifferential operators and stabilization of solutions of parabolic equations with random coefficients, *Mat. Sb.* **1113**, no. 1 (1980), 41-64 (in Russian).
17. Dixmier, J.: C^**-algebras*, North-Holland, 1987 (transl. from the French).
18. Diestel, J.: *Geometry of Banach Spaces*, Springer, 1975.
19. Dubinskii, Yu.A.: Weak convergence in nonlinear elliptic and parabolic equations, *Mat. Sb.* **67**, no. 4 (1965), 609-642 (in Russian).

20. Dubrovin, B.A., Matveev, V.B., and Novikov, S.P.: Nonlinear equations of Korteweg-de Vries type, finitely-zoned operators and Abelian varieties, *Uspekhi Mat. Nauk.* **31**, no. 1 (1976), 55-136 (in Russian).
21. Duvant, G. and Lions, J.-L.: *Inequalities in Mechanics and Physics*, Springer, 1975 (transl. from the French).
22. Zhikov, V.V.: Monotonicity in the theory of almost periodic solutions of nonlinear operator equations, *Mat. Sb.* **90**, no. 2 (1973), 214-227 (in Russian).
23. Zhikov, V.V.: On the solvability of nonlinear equations in the classes of Bohr and Besicovitch almost periodic functions, *Mat. Zam.* **18**, no. 4 (1975), 553-560 (in Russian).
24. Zhikov, V.V.: Some questions on admissibility and dichotomy. Averaging principle, *Izv. Akad. Nauk SSSR Ser. Mat.* **40**, no. 6 (1976), 1380-1408 (in Russian).
25. Zhikov, V.V. and Levitan, B.M.: Favard theory, *Uspekhi Mat. Nauk* **32**, no. 2 (1977), 123-171 (in Russian).
26. Kadets, M.I. and Kyursten, K.: Countability of the spectrum of an almost periodic function with values in a Banach space. In: *Theory of Functions, Functional Analysis and their Applications*, Kharkov Univ. **33** (1980), 45-49 (in Russian).
27. Kachurovski, R.I.: Nonlinear monotone operators in Banach spaces, *Uspekhi Mat. Nauk.* **23**, no. 2 (1968), 121-168 (in Russian).
28. Kelley, J.L.: *General Topology*, Springer, 1975.
29. Kiselev, V.Yu.: Almost periodic Fourier integral operators and their applications, *Tr. Sem. Petrovskii* **3** (1977), 81-97 (in Russian).
30. Kiselev, V.Yu.: Almost periodic Fourier integral operators and hyperbolic equations in spaces of almost periodic functions, *Dokl. Akad. Nauk SSSR* **244**, no. 1 (1979), 33-37 (in Russian).
31. Kiselev, V.Yu.: Strongly hyperbolic equations in spaces of almost periodic functions, *Tr. Sem. Petrovskii* **7** (1981), 125-147 (in Russian).
31a. Kochubei, A.N.: On almost periodic solutions of operator differential equations, *Sib. Mat. Zh.* **24**, no. 3 (1083), 102-111 (in Russian).
32. Krasnosel'skii, M.A.: *Topological Methods in the Theory of Nonlinear Integral Equations*, Pergamon, 1964, (transl. from the Russian).
33. Krasnosel'skii, M.A. and Rutitskii, Ya.B.: *Convex Functions and Orlicz Spaces*, Atomic Energy Commission USA, 1960 (transl. from the Russian).
34. Krasnosel'skii, M.A., Rutitskii, Ya.B., and Sultanov, R.M.: On a nonlinear operator acting in a space of abstract functions, *Izv. Akad. Nauk SSSR Ser. Mat. Fiz. i Tekhn.* **3** (1959), 15-21 (in Russian).
35. Krasnosel'skii, M.A., Burd, V.Sh., and Kolesov, Yu.S.: *Nonlinear Almost*

Periodic Oscillations, Wiley, 1973 (transl. from the Russian).
36. Krein, S.G.: *Linear Differential Equations in Banach space*, Amer. Math. Soc., 1971 (transl. from the Russian).
37. Krein, S.G.: *Linear Equations in Banach Spaces*, Birkhäuser, 1982 (transl. from the Russian).
38. Krein, S.G., Petunin, Yu.I., and Semenov, E.M.: *Interpolation of Linear Operators*, Amer. Math. Soc., 1982.
39. Courant, R.:. *Partial Differential Equations*, Interscience, 1965.
40. Levitan, B.M.: *Almost periodic Functions*, Gostekhizdat Moscow, 1953 (in Russian).
41. Levitan, B.M. and Zhikov, V.V.: *Almost Periodic Functions and Differential Equations*, Cambridge Univ. Press, 1982 (transl. from the Russian).
42. Lang, S.: *Algebra*, Addison-Wesley, 1984.
43. Lions, J.L.: On inequalities in partial derivatives, *Uspekhi Mat. Nauk.* **26**, no. 2 (1971), 205-263 (in Russian).
44. Lions, J.L.: *Quelques Méthodes de Résolution des Problèmes aux Limites non-Linéaires*, Dunod, 1969.
45. Lions, J.L.: *Optimal Control of Systems Governed by Partial Differential Equations*, Springer, 1971 (transl. from the French).
46. Lions, J.L. and Magènes, E.: *Non-homogeneous Boundary Value Problems and Applications*, Springer, 1972, 2 Vols. (transl. from the French).
47. Mizohata, S.: *The Theory of Partial Differential Equations*, Cambridge Univ. Press, 1973 (transl. from the Japanese).
48. Moseenkov, V.B.: Almost periodic solutions of nonlinear wave equations, *Ukr. Mat. Zh.* **29**, no. 1 (1977), 112-118 (in Russian).
49. Moseenkov, V.B.: Application of the method of accelerated convergence to the study of a nonlinear wave equation, *Ukr. Mat. Zh.* **30**, no. 1 (1978), 54-62 (in Russian).
50. Mukhamadiev, E.M.: On invertibility of partial differential operators of elliptic type, *Dokl. Akad. Nauk SSSR* **205**, no. 6 (1972), 1292-1295 (in Russian).
51. Narasimhan, R.: *Analysis on Real and Complex Manifolds*, Springer, 1971.
52. Nirenberg, L.: *Lectures on Linear Partial Differential Equations*, Amer. Math. Soc., 1973.
53. Pankov, A.A.: On almost periodic linear operators with parameter, *Soobshch. Akad. Nauk. GruzSSR* **87**, no. 2 (1977), 293-295 (in Russian).
54. Pankov, A.A.: On almost periodic solutions of evolution variational inequalities, *Dokl. Akad. Nauk* **241**, no. 2 (1978), 286-289 (in Russian).
55. Pankov, A.A.: On almost periodic solutions of evolution equations, *Diff. Eq.* **14**, no. 6 (1978), 1140- 1143 (in Russian).

56. Pankov, A.A.: Discrete approximations of convex sets and convergence of solutions of variational inequalities, *Math. Nachr.* **91** (1979), 7-22 (in Russian).
57. Pankov, A.A.: Bounded and almost periodic solutions of evolution variational inequalities, *Mat. Sb.* **108**, no. 4 (1979), 551-556 (in Russian).
58. Panov, A.A.: Regularity of solutions of abstract variational inequalities, *Ukr. Mat. Zh.* **32**, no. 5 (1980), 683-686 (in Russian).
59. Pankov, A.A.: On Favard's theory for impulse evolution equations, *Rev. Roum. Math. Pures Appl*, **25**, no. 3 (1980), 385-401 (in Russian).
60. Pankov, A.A.: On Besicovitch almost periodic solutions of evolution variational equations, *Ukr. Mat. Zh.* **33**, no. 1 (1981), 91-96 (in Russian).
61. Pankov, A.A.: On the theory of almost periodic pseudodifferential operators of order 1, *Ukr. Mat. Zh.* **33**, no. 5 (1981), 615-619 (in Russian).
62. Pankov, A.A.: Almost periodic solutions of the nonlinear Schrödinger equation, *Mat. Met. i Fiz.- Mekh. Pol.* **14** (1981), 27-30 (in Russian).
63. Pankov, A.A.: On nonlinear monotone partial differential operators with almost periodic coefficients, *Dokl. Akad. Nauk. Ukr. SSR Ser.* A **5** (1981), 20-22 (in Russian).
64. Panov, A.A.: On nonlinear hyperbolic systems with almost periodic coefficients, *Dokl. Akad. Nauk. Ukr. SSR Ser.* A **8** (1981), 32-35 (in Russian).
65. Pankov, A.A.: Invertibility of partial differential operators of order 1 with almost periodic coefficients. In: *Boundary Value Problems of Mathematical Physics*, Nauk. Dumka Kiev, 1981, 102-103 (in Russian).
66. Pankov, A.A.: On bounded and almost periodic solutions of certain nonlinear evolution equations, *Uspekhi Mat. Nauk*, **37**, no. 2 (1982), 223-224 (in Russian).
67. Pankov, A.A.: Boundedness and almost periodicity in time of solutions of evolution variational inequalities, *Izv. Akad. Nauk SSSR Ser. Mat.* **46**, no. 2 (1982), 314-346 (in Russian).
68. Pankov, A.A.: Bounded and almost periodic in time solutions of a class of nonlinear evolution equations, *Mat. Sb.* **121**, no. 1 (1983), 72-86 (in Russian).
69. Pankov, A.A.: On nonlinear elliptic equations of order two with almost periodic coefficients, *Ukr. Mat. Zh.* **35**, no. 5 (1983), 649-652 (in Russian).
70. Pastur, L.A.: Spectra of random self-adjoint operators, *Uspekhi Mat. Nauk.* **28**, no. 1 (1973), 3-64 (in Russian).
71. Perov, A.I. and Trubnikov, Yu.B.: Monotone differential equations, II, *Diff. Eq.* **12**, no. 7 (1976), 1223-1237 (in Russian).
72. Pikulin, V.P.: On periodic and almost periodic solutions for a class of quasilinear parabolic equations, *Diff. Eq.* **18**, no. 8 (1982), 1412-1417 (in Russian).

73. Pontryagin, L.S.: *Topological Groups*, Princeton Univ. Press, 1958 (transl. from the Russian).
74. Reed, M., and Simon, B.: *Methods of Modern Mathematical Physics*, Vol. 4, Acad. Press, 1978.
75. Rokhlin, V.A. and Fuks, D.B.: *Beginner's Course in Topology. Geometric Chapters*, Springer, 1984 (transl. from the Russian).
76. Skrypnik, I.V.: *Nonlinear Elliptic Equations of Higher Order*, Nauk. Dumka, Kiev, 1973 (in Russian).
77. Slyusarchuk, V.E.: Bounded solutions of nonlinear elliptic equations, *Uspekhi Mat. Nauk* **35**, no. 1 (1980), 215-216 (in Russian).
78. Sobolev, S.L.: On almost periodic solutions of the wave equation, *Dokl. Akad. Nauk SSSR* **48**, no. 8 (1945), 570-574; **48**, no. 9 (1945), 646-650; **49**, no. 1 (1945), 12-15 (in Russian).
79. Sobolev, S.L.: *Applications of Functional Analysis in Mathematical Physics*, Amer. Math. Soc, 1973 (transl. from the Russian).
80. Stein, E.M.: *Singular Integrals and Differentiability Properties of Functions*, Princeton Univ. Press, 1970.
81. Stein, E.M., and Weiss, G.: *Fourier Analysis on Euclidean Spaces*, Princeton Univ. Press, 1971.
82. Tikhonov, A.N.: Systems of differential equations having a small parameter in front of the derivatives, *Mat. Sb.* **31**, no. 3 (1952), 576-586 (in Russian).
83. Triebel, H.: *Interpolation Theory, Function Spaces, Differential Operators*, North-Holland, 1978.
83a. Halanay, A. and Wexler, D.: *Teoria Calitative a Sistemelor cu Impulsuri*, Ed. Acad. Romania, 1968.
84. Hartman, P.: *Ordinary Differential Equations*, Birkhäuser, 1982.
85. Heyer, H.: *Probability Measures on Locally Compact Groups*, Springer, 1977.
86. Hayman, W.K.: *Meromorphic Functions*, Clarendon Press, 1964.
87. Hewitt, E. and Ross, K.: *Abstract Harmonic Analysis*, Springer, 1979, 2 Vols.
88. Schaefer, H.H.: *Topological Vector Spaces*, MacMillan, 1966.
89. Shubin, M.A.: Differential and pseudodifferential operators in spaces of almost periodic functions, *Mat. Sb.* **95**, no. 4 (1974), 560-584 (in Russian).
90. Shubin, M.A.: Theorems on the coincidence of the spectra of pseudodifferential operators in the spaces $L^2(\mathbf{R}^n)$ and $B^2(\mathbf{R}^n)$, *Sib. Mat. Zh.* **17**, no. 1, (1976), 200-215 (in Russian).
91. Shubin, M.A.: Almost periodic functions and partial differential operators, *Uspekhi Mat. Nauk.* **33**, no. 2, (1978), 3-47 (in Russian).
92. Shubin, M.A.: *Pseudodifferential Operators and Spectral Theory*, Springer, 1987 (transl. from the Russian).

93. Shubin, M.A.: Local Favard theory, *Vestnik Moskov. Univ. Ser.* 1, **2** (1979), 31-36 (in Russian).
94. Shubin, M.A.: Spectral theory and the index of elliptic operators with almost periodic coefficients, *Uspekhi Mat. Nauk* **34**, no. 2 (1979), 95-135 (in Russian).
95. Edwards, R.: *Functional Analysis*, Holt, Rinehart & Winston, 1965.
96. Ekeland, I. and Temam, R.: *Analysis Convexe et Problèmes Variationelles*, Dunod, 1974.
97. Agmon, S.: The L^p-approach to the Dirichlet problem, *Ann. Scu. Norm. Sup. Pisa* **13**, no. 4 (1959), 405-448.
98. Amerio, L. : Soluzioni quasi-periodiche, o limitate, di sistemi differenziali non-lineari quasi-periodiche, o limitate. *Ann. Mat. Pura. Appl* **39**, (1955), 97- 119.
99. Amerio, L.: Funzioni debolmente quasi-periodiche. *Rend. Sem. Mat. Univ. Padova* **30** (1960), 288- 301.
100. Amerio, L., and Prouse, F.: *Almost periodic funcitions and functional equations* Van Nostrand, 1971.
101. Artola, M.: Sur les perturbations des èquations d'èvolution. Application à problèmes de retard, *Ann. Sci. Ecole Norm. Sup*, Paris **2** (1969), 137-153
102. Asplund, E.: Averaged normes, *Israel. J. Math* **5** (1967), 227-233.
103. Attouch, H., and Damlamjan, A. : Problems d'évolution dans hilberts et applications, *J. Math. Pures Appl* **54**, no. 1 (1975), 53-74.
104. Bardos, C., and Brézis, H.: Sur une classe de problemes d'évolution non linearies, *J. Diff. Eq.* **9** (1969), 345-394.
105. Besicovitch, A., and Bohr, H.: Almost periodicity and generalized trigonometric series, *Acta Math* **57** (1931), 203-291.
106. Besicovitch, A.: *Almost periodic functions*, Dover, 1954
107. Biroli, M.: Sull'unicita della soluzione limitate di una disequaziona variazionale d'evoluzione, *Rend. Accad. Naz. Lincei Ser.* 8 **48**, no. 4 (1970), 409-411.
108. Biroli, M.: Solutions presque pèriodiques des inéquations d'évolutions avec fonctionelles non differentiables, *Rend. Sem. Mat. Univ. Padova* **44** (1970), 299-318.
109. Biroli, M.: Sur les solutions presque pèriodiques des inéquations d'évolution paraboliques. *Ann. Mat. Pura Appl.*, **88** (1971), 51-70.
110. Biroli, M.: Sur les solutions bornées ou presque periodiques des équations d'évolution multivoques sur un espace de Hilbert, *Ricerche Mat.* **21** (1972), 17-47.
111. Biroli, M.: Sur les solutions bornées presque périodiques des équations et inéequations d'évolution. *Ann. Mat. Pura. Appl*, **93** (1972), 1-79.
112. Biroli, M.: Sur l'inéquations d'évolution de Navier. Stokes I, II, III, *Rend.*

Accad. Naz. Lincei. Ser. 8, **52** (1982), 457-460; 591-598; 811-820.

113. Bochner, S.: Abstrakte Fastperiodische Funktionen. *Acta. Math.* **61** (1933), 149-184.
114. Bochner, S.: Fastperiodische Losungen der Wellengleichung, *Acta. Math.* **62** (1933), 227-237.
115. Bohl, P.: Ueber eine Differentialgleichung der Störungstheorie, *Crelles J.* **131** (1906), 268-321.
116. Bohr, H.: Zur Theorie der fastperiodischen Funktionen, I, II, *Acta. Math.*, **45** (1925), 29-127; **46**, (1927), 101-214.
117. Brézis, H.: Équations et inéquations non lineaires dans les espaces vectoriells en dualite, *Ann. Inst. Fourier*, **18** (1968), 115-175.
118. Brézis, H.: Problems unilateraux. *J. Math. Pures Appl.* **51** (1972), 1-168.
119. Brézis, H.: *Operateurs maximaux monotones et semigroups de contractions dans les espaces de Hilbert*, North Holland, 1973.
120. Brézis, H., Grandall, M., and Pazy, A.: Perturbations of non linear maximal monotone sets in Banach spaces, *Comm. Pure Appl. Math.* **23**, no. 1 (1972), 123-144.
121. Brézis, and Nirenberg, L.: Some first order non linear equations on a torus, *Comm. Pure Appl. Math.* **30**, no. 1 (1977), 1-11.
122. Brézis, H., and Nirenberg, L.: Characterization of the range of some nonlinear operators and applications to boundary value problems, *Ann. Scu. Norm. Sup. Pisa*, **5**, no. 2 (1978), 225-326.
123. Bréisz, H., and Nirenberg, L.: Firced vibrations for a nonlinear wave equation, *Comm. Pure Appl. Math.* **31**, no. 1 (1978), 1-30.
124. Browder, F., and Hess, P.: Nonlinear mappings of monotone type in Banach spaces. *J. Funct. Anal.* **11** (1972), 251-294.
125. Calderon, A.P.: Existence and uniqueness theorems for systems of partial differential equations, in: *Proc. Symp. Fluid Dynamics and Appl.* Univ. Maryland, 1961 Gordon and Breach, 1962, 147-195.
126. Favard, J.: Sur les équations différentielles a coèfficients presque périodiques, *Ann. Math.* **51** (1927), 31-81.
127. Fink, A.M.,: Almost periodic differential equations *Lect. Notes Math.* **377**, Springer, 1974.
128. Friedrichs, K.: Symmetric hyperbolic systems of linear differential operators, *Comm. Pure Appl. Math.*, 7 (1954), 345-392.
129. Friedrichs, K.O., and Lax, P.D.: Boundary value problems for first order operators, *Comm. Pure Appl. Math.*, **18** (1965), 365-388.
130. Friedrichs, K.O., and Lax, P.D.: On symetrizable differential operators, *Proc. Symp. Pure Math.*, **10**, Amer. Math. Soc., 1968, 128-137.

131. Grothendieck, A.: Produits tensoriells topologiques et espaces nucléaires, *Mem. Amer. Math. Soc.* **16**, Amer. Math. Soc., 1955.
131a. Haraux, A.: Nonlinear Evolution Equations. Global Behavior of Solutions, *Lect. Notes Math.* **841**, Springer, 1981.
132. Huet, D.: Perturbations singuliéres d'inéquations variationelles, *Compt. Rend. Acad. Sci. Paris*, **267** (1968), 932-934.
133. Itoh, S. : Nonlinear random equations with monotone operators in Banach spaces, *Math. Ann.*, **236**, no. 2 (1978), 133-146.
134. Itoh, S.: Random differential equations associated with accretive operators, *J. Diff. Eq.*, **31**, no. 1 (1979), 139-154.
135. Ka-Sing, Lau.: On the Banach spaces of functions with bounded upper means, *Pacific J. Math.*, **91**, no. 1 (1980), 153-172.
136. Kravvaritis, D.: Nonlinear random operators of monotone type in Banach spaces, *J. Math. Anal. Appl.*, **78**, no. 2 (1980), 488-496.
137. Kravvaritis, D.: Existence theorems for nonlinear random equations and inequalities, *J. Math. Anal. Appl.*, **88**, No. 1 (1982), 61-75.
138. Leray, J., and Lions, J.-L.: Quélques resultats de Visik sur les problémes elliptiques non linéaires par les méthodes de Minty-Browder, *Bull. Soc. Math. France*, **93** (1965), 97-107.
139. Lions, J.-L.: Sur certains systémes hyperboliques non linéaires, *Compt. Rend. Acad. Sci. Paris*, **257** (1963), 2057-2060.
140. Lions, J.-L., and Stampacchia, G.: Variational inequalities, *Comm. Pure Appl. Math.*, **20** (1967), no. 3, 493-519.
141. Marcinkiewicz, J.: Une remarque sur les espaces de Besicovitch, *Compt. Rend. Acad. Sci. Paris*, **208** (1939), 152-159.
142. Mosco, U.: Convergence of convex sets and of solutions of variational inequalities, *Adv. Math.*, **3**, No. 4 (1969), 510-585.
143. Pozzi, G.: Sulle equazioni astratte lineare e nonlineare del tipo di Schrödinger. *Ann. Mat. Pura Appl.*, **78** (1968), 197-258.
144. Prouse, G.: Soluzioni quasi-periodiche dell'equazione di Navier-Stokes in due dimensioni, *Rend. Sem. Mat. Univ. Padova*, **33** (1963), 186-212.
145. Prouse, G.: Periodic or almost periodic solutions of a non-linear functional equation, *Rend. Accad. Naz. Lincei, Ser. 8*, **43** (1967), 161-167; 281-287; 448-452; **44** (1968), 1-10.
146. Rabinowitz, P.H.: Free vibrations for a semilinear wave equation, *Comm. Pure Appl. Math*, **31**, no. 1 (1978), 31-68.
147. Tartar, L.: Interpolation non-linéaire et regularité, *J. Funct. Anal.*, **9** (1972), 469-489.
148. Zaidman, S.: Solutions presque périodiques des équations différentielles abstraites, *l'Ennseignem. Math.*, **24** (1978), no. 1-2, 87-110.

Subject Index